Real Analysis

Alice Gorguis

Mathematics Department, North Park University,

Chicago, IL. 60625

3rd Edition

To order additional copies of this book, contact:
Xlibris
1-888-795-4274
www.Xlibris.com
Orders@Xlibris.com
721369

Dedication:

This book is dedicated to my students of fall'2011 at North Park University especially, Zechariah Gelzer, and Tyler Swanson, who inspired me to write the first edition of this book in 2011.

You have heard me teach many things that have been confirmed by many reliable Witnesses. Teach these great truths to trustworthy people who are able to pass them on to others.

Timothy 2 : 2 $p1338$

Preface:

This edition is a modification for my second edition of " Real Analysis-Step-by-Step Approach" that was published on Spring 2011. The book is designed for students who have completed the ordinary course in elementary calculus and Discrete Mathematics, and it covers a portion of the material that every student in mathematics is familiar with. I hope that this book can enable the student to learn enough examples, theorems, and methods in analysis.

It has been the purpose to limit the amount of material to what can readily be covered in a semester of four hours per week, and to select those topics for discussion which the student might well know who expects to teach Calculus or to continue his/her study of Analysis. The material presented represents a departure from that usually given in "Calculus" courses. It represents introduction to the fundamentals of real analysis including real numbers, limits, sequences, derivatives, the Riemann integrals and metric spaces. The chapters begin with the important tools used in analysis: the axiom of the real number system, distance and their properties, sequence and their limits.

My goal of writing this book is to help the students to learn the style of thinking, and learn the basic ideas and techniques of analysis, for functions of a single real variable, in an easy way. I include some exercises and examples that help to connect the abstract concepts being discussed to the student's own understanding of more concrete ideas.

Methods of Teaching:

A typical class will be a mixture of student presentations and discussion. Students will be given a reading assignments; proofs; and projects to work on.

Note for the students:
Students will be confronted with a statement to prove, that can be challenging business, here is a simple hint to follow: You should first replace terms by their definition and then carefully analyze what the hypothesis and the conclusion mean. After doing so, you can attempt to prove the result using one of the available methods of proof.

Finally, I wish to acknowledge here my indebtness to my students of spring' 2011 especially Zechariah Gelzer and Tyler Swanson, for inspiring me to write the first edition, and I also like to mention in particular my two students who read the second editions and gave the following comments on the book:

Stephen George wrote: "A Simpler Approach to Real Analysis" is effective in helping users build a strong foundation of mathematical thinking. The concepts are logically organized and move at a manageable pace, which supplemented with a professor's guidance, break down theoretical math to a level that even a slow learner like me can understand. This textbook is especially handy for reference as it avoids being needlessly wordy and is packed with practical examples.

John Fredrickson wrote: Having little experience in higher levels of math before taking this class, A Simpler Approach to Real Analysis proved to be a very helpful and well-organized book. Clear examples and well-written proofs helped make the subject easier to understand, even to someone who had taken a break from higher-level math courses for a few years. The practice problems were explained clearly, especially in relation to the concepts they required, and helped clarify the subjects that they required. I'm glad I had this book to help me through my course in Real Analysis.

This third edition differs from the second one in the followings:
Additional explanation, examples and graphs are added to most of the sections, and addition of theories from calculus

A. Gorguis
June, 2015

Contents

CONTENTS

Chapter 1

The Real Numbers

$\mathbf{R}$eal numbers are considered to be the most fundamental mathematical structure. One of its most important sets is the set R, whose elements are the real numbers.

Every real number can be represented by a point on a line called the real line ,or the line of real numbers, and conversely. The set of real numbers is the set of numbers that are rational or irrational:

$$R = \{x \mid x \text{ is rational or irrational}\} \qquad (1.1)$$

Where the rational numbers can be expressed as an integer divided by a nonzero integer. The set of rational numbers are:

$$\{\frac{a}{b} \mid a \text{ and } b \text{ are integers and } b \neq 0\}.$$

And the set of irrational number is the set of numbers who's decimal representations neither terminates nor repeat. Irrational numbers can not be expressed as quotients of integers. Set of real numbers includes:

Open intervals in R such as, $\{x \mid a < x < b\}$, or (a, b).

For example $\{x|2 < x < 5\}$ means $x =$ set of elements 3, and 4 or $x = \{3, 4\}$ as shown on the line of numbers.

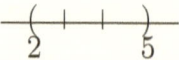

Closed intervals in R such as, $\{x|a \le x \le b\}$ or, $[a, b]$.
For example $\{x|2 \le x \le 5\}$ means $x =$ set of elements 2, 3, 4 and 5 or $x = \{2, 3, 4, 5\}$ as shown on the line of numbers.

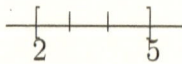

Or half open , half closed intervals in R such as $\{x|a < x \le b\}$ or $\{x|a \le x < b\}$, denoted by $(a, b], [a, b)$ respectively.
For example $\{x|2 < x \le 5\}$ means $x =$ set of elements 3, 4 and 5 or $x = \{3, 4, 5\}$,or $\{x|2 \le x < 5\}$ means $x =$ set of elements 2, 3, and 4 or $x = \{2, 3, 4\}$ as shown , respectively, on the line of numbers.

The set of real numbers can be extended to include $\mp\infty$, thus $\{x : -\infty < x < \infty\}$ or, $(-\infty, \infty)$ in R.

Types of Real Numbers used are:

a. Natural Numbers:
They are the usual counting numbers e.g 1, 2, 3, .. or as a set $\{1, 2, 3, ..\}$.

b. Whole Numbers:
They are the Natural numbers with addition of zero, as a set $\{0, 1, 2, 3, ..\}$.

c. Integers:
Consists of the natural numbers, 1, 2, 3, .. , the number zero, and the negative numbers, or in set notation:
$\{... -3, -2, -1, 0, 1, 2, 3, ...\}$ or simply stated as $\{0, \mp\}$.

Example: Consider a natural number 2, we will built other numbers such as:

$$2 + 3 = 5.$$
$$2 + 2 = 4.$$
$$2 + 1 = 3.$$
$$2 + 0 = 2.$$
$$2 - 1 = 1.$$
$$2 - 2 = 0.$$
$$2 - 3 = -1.$$
$$2 - 4 = -2.$$
$$2 - 5 = -3.$$
$$2 - 6 = -4.$$
$$2 - 7 = -5.$$

Notice the order of "greater than", and "less than", which form the interesting pattern on the line of numbers:

d. Rational Numbers:
Rational numbers or fractions, are numbers of the quotient $\frac{p}{q}$, where both p, and q are integers. assuming that $q \neq 0$. For example $\frac{3}{7}$, and $\frac{-2}{11}$ are rational numbers. We can write the set of rational numbers on the line of numbers as:

$$\{\frac{-1}{1}, \frac{-3}{4}, \frac{-2}{3}, \frac{-1}{2}, \frac{-1}{3}, \frac{-1}{4}, \frac{0}{1}, \frac{1}{4}, \frac{1}{3}, \frac{1}{3}, \frac{1}{2}, \frac{2}{3}, \frac{3}{4}, \frac{1}{1}, \frac{5}{4}, \frac{4}{3}, \frac{3}{2}, \frac{5}{3}, \frac{7}{4}...\}.$$

The set of rational numbers with the two basic operations of addition and multiplication form a structure called an **ordered field** where the following properties are true:

a) closed   b) commutative   c) associative

4. There are two numbers: 0, and 1, called identities such that such that for any real number a :

a) $0 + a = a$

b) $1.a = a$.

5. Multiplication is distinctive over addition.

6. The set of rational numbers can be separated into 3-distinct subsets: a) Negative numbers   b) Zero   c) Positive numbers.

Example: If a pie is cut into 8-equal portions, then the idea of comparing two quantities is called portion i.e the portion of pie to the whole pie is 1 to *eight*, and we write:

$$\frac{portion}{whole\ pie} = \frac{1}{8}.$$

$$Rational\ numbers\ or\ fractions\ P = \frac{1}{8}.$$

Thus, the rational number is the quotient of two integers.

$$\frac{Numerator}{Denominator} = \frac{N}{D} = R\ the\ rational\ number, D \neq 0.$$

e. Irrational Numbers:

Irrational numbers are the real numbers, such as $\sqrt{2}$, π , and e, that are not rational numbers.

Example: Let $abcd$ be a unit square,

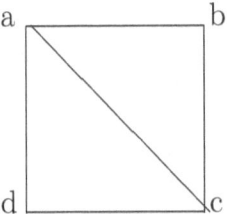

Using Pythagorean Theorem we have:

$$\overline{ac}^2 = \overline{ad}^2 + \overline{dc}^2.$$
$$\overline{ac}^2 = 1^2 + 1^2.$$
$$Or \ \overline{ac} = \sqrt{2}.$$

Which is not a rational number.
If we consider $\sqrt{2}$ to be a rational number, then we can write:

$$\sqrt{2} = \frac{p}{q}.$$
$$Squaring \ both \ sides \ gives \ 2 - (\frac{p}{q})^2 = \frac{p^2}{q^2}.$$
$$Then, \ 2q^2 = p^2.$$

i.e we must have $p = 2C$, where C is an integer, this gives $2q^2 = (2C)^2$ or $q^2 = 2C^2$.
But this leads us to the conclusion that both p, and q must be even which contradicts our initial statement that $\frac{p}{q}$ is a rational number. Thus we must conclude that $\sqrt{2}$ is not a rational number. Using a simple division, we can approximate $\sqrt{2}$ to any desired decimal places:

$$1.4 < \sqrt{2} < 1.5.$$
$$1.41 < \sqrt{2} < 1.42.$$
$$1.414 < \sqrt{2} < 1.415.$$
$$1.4142 < \sqrt{2} < 1.4143.$$
$$1.41421 < \sqrt{2} < 1.41422.$$
$$1.414214 < \sqrt{2} < 1.414215.$$

.

.

Here, we give a list of set notations that play an important role in discrete mathematics:

N = {0, 1, 2, 3, . . . }, the set of natural(whole) numbers.

I = {. ., -2, -1, 0, 1, 2, . .},the set of integers.

I^+ = {1, 2, 3, . . }, the set of positive integers.

Q = {p/q: p∈ I, q∈I, and q≠ 0}, the set of rational numbers.

R = {set of real numbers}.

$\mathbf{R}^+$= {set of positive real numbers }.

C = { set of complex numbers }.

Order in Real Numbers: The real number system begins with a set R, Which contains elements called the real numbers. The set R with addition, and multiplication has properties, that are sufficient to determine the computational rules involved in factoring, working with fractions, and exponents, and solving simple algebraic equations, which we are familiar with. The results derived from these properties deal with equality or identity of two expressions. Dealing with real numbers, a unique subset p of real number R called the set of positive real numbers, such that the following two conditions are satisfied:

1. For each real number a exactly one of the following statements holds:

i) a is an element of p.

ii) $a = 0$.

iii) $-a$ is an element of p, where $p \subset R$.

$$
\begin{array}{ccc}
-a & a = 0 & +a
\end{array}
$$

line of numbers ⎯⎯⎯⎯⎯⎯⎯⎯⎯⎯⎯⎯

2. If a and b are any positive numbers, then $a + b$ is positive real number, and $a\,b$ is a positive real number too.

1.1 Properties of Real Numbers

Let the set of real numbers be given as: $R = \{a, b, c, ...\}$ on which the operation of additions and multiplications are defined so that every pair of real numbers has a unique sum and product, and has the following properties:

1. $a + b = b + a$ and $ab = ba$ (commutative laws).

2. $(a+b)+c = a+(b+c)$ and $(ab)c = a(bc)$, (associative laws).

3. $a(b + c) = ab + ac$ (distributive laws).
For example : The relations of binomial,

$$
\begin{align}
(a + b)^2 &= a^2 + 2ab + b^2, & (1.2) \\
(a + b)(c + d) &= ac + ad + bc + bd, & (1.3) \\
(-a) &= (-1)a, & (1.4) \\
a(-b) &= (-a)(b) = -ab, & (1.5) \\
and \quad \frac{a}{b} + \frac{c}{d} &= \frac{ad + cb}{bd}, \ (b, d \neq 0). & (1.6)
\end{align}
$$

Follows from properties $(1 - 3)$.

4. There are distinct real numbers $0, 1$, such that:
$a + 0 = a$, and $a.1 = a$ for all a.

5. For each a there is a real number $(-a)$ such that
$a + (-a) = 0$, and if $a \neq 0$ there is a real number $\frac{1}{a}$ such that:
$a(\frac{1}{a}) = 1$.

1.2   Proofs

Since real analysis deals with many proofs, then we have to introduce the notion of proof. A proof is a valid argument that establishes the truth of a mathematical statement. A proof can use the hypotheses of the theorem, if any, axioms assumed to be true, and previously proven theorem. Proofs are extremely important for computer science in verifying that a computer programs are correct.

Terminology used:
• Theorem: is a statement that can be proofed to be true. Theorem can also be referred to as: Facts, or Results.
• Proposition: Less important that theorem.
• Axiom , or postulates: Statements assumed to be true.
• Lemma: Proofs of other results.
• Corollary: is a theorem that can be established directly from a proven theorem.
• Conjecture: is a statement that is being proposed to be a true statement.

Methods of Proving: Proof can be done directly, indirectly, or by contradiction.

Example-1: Direct Proof
Prove that $(n + 1)^2 \geq 2^n$, if n is a positive integer with $n \leq 4$.

Solution: With direct substitution, we can verify the inequality as follows:

for $n = 1 \rightarrow 2^2 = 4 \geq 2^1 = 2$.

for $n = 2 \rightarrow 3^2 = 9 \geq 2^2 = 4$.

for $n = 3 \rightarrow 4^2 = 16 \geq 2^3 = 8$.

for $n = 4 \rightarrow 5^2 = 25 \geq 2^4 = 16$.

In all these cases the theorem is proved for positive n.

Example-2: Indirect Proof

Prove that if n is an integer and n^2 is an even, then n is even.

Solution: Since n is an integer, then n is either even or odd, suppose n is even then there exist an integer m such that :
$n = 2m \rightarrow n^2 = 4m^2 = 2(2m)$ then n is even.

Example-3: Proof by Contradiction

Proof that if $3n + 1$ is odd, then n is odd.

Solution: Using logic, let "$p = 3n + 1$ is odd", and "$q = n$ is not odd (contradiction)". so we have the following truth value:

if $p \rightarrow \sim q = T \rightarrow \sim T = T \rightarrow F = F$ Then the original statement holds.

1.2.1 Graphs of Sets of Real Numbers

<u>Definition</u>: Let A be a set of ordered pairs of real numbers. The graph of A is the set of all points in the plane whose coordinates (ordered pairs of real numbers) are elements of A.

Example-1: What is the graph of the set:

$$C_1 = \{(a, b); a^2 + b^2 = 1\}.$$

Solution: The point $(a, b) \in S_1$ IFF,

$$\sqrt{(a - 1)^2 + (b - 1)^2} = 1.$$

Which means that (a, b) is a point whose distance from the origin is $1 \Rightarrow$ the graph of C_1 is the set of all points at distance of 1-unit from the origin. Then this is a circle centered at the origin with radius $= 1$.

Example-2: In a similar way as in Example-1 we can show that the graph of the set $C_r = \{(a, b); a^2 + b^2 = r^2\}$, where $r > 0$, is the graph of a circle centered at the origin , but with a radius $= r$, with $r > 0$.

Example-3: What is the graph of (x, y) for :

$$(x + 3)^2 + (y - 7)^2 = 4.$$

By following the steps from examples 1, and 2, student can easily show that this graph is a graph of a circle centered at $(-3, 7)$ and radius $r = 2$.

1.3   The Real Number Axioms

Consider a set of real numbers R, the set of positive real numbers P, and functions '+' (addition), and '.' (multiplication)on $R \times R \Rightarrow R$ and assume they satisfy the following axioms:

1. The field axioms: This group describes the algebraic properties for all real numbers $x, y, and\ z$, and any field that follows the following axioms is called: a field under , *addition*, *multiplication*. If x, y, and z are any numbers then,
a) $x + y = y + x$ (Commutative Law).

b) $(x + y) + z = x + (y + z)$(Associative Law).

c) $0 \in R$ such that $x + 0 = x$ for all $x \in R$.

d)   for each $x \in R$ there is a $-x \in R$ such that $x + (-x) = 0$.

e) $xy = yx$.

f)   there is $x \in R$ such that $x.0 = 0$,and $x.1 = x$ for all $x \in R$.

g)   for each $x \in R$ different from 0 there is $w \in R$ such that $xw = 1$.

h) $x(y + z) = xy + xz$. Distributive Property.

2.   The order axioms: This axiom is due to the fact that real numbers are ordered. The set P of positive real numbers satisfy the following:

a) $x, y \in P \Rightarrow x + y \in P$.

b) $x, y \in P \Rightarrow xy \in P$.

c) $x \in P \Rightarrow -x \notin P$.

d) $x \in R \Rightarrow (x = 0) or (x \in P) or (-x \in P)$.

e) $x, y \in R \Rightarrow x - y = x + (-y)$.

f) $x, y \in R \Rightarrow \frac{x}{y} = x.\frac{1}{y} \in R$.

Any system satisfying the axioms of group (1), and(2) is called an Ordered Field. In an ordered field we describe the following notations:

$x < y$ means $y - x \in P$, or $x < y$ if and only if $y - x$ is positive;

$x > y$ means $y - x \in P$, or $x > y$ if and only if $x - y$ is positive;

$x \leq y$ means $x < y$ or $x = y$,or $x \leq y$ if and only if either $x < y$ or $x = y$;

$x \geq y$ means $x > y$ or $x = y$,or $x \geq y$ if and only if either $x > y$ or $x = y$.

Also, from (2) axiom (a) is equivalent to : $(x < y \ \& \ z < w) \Rightarrow x + z < y + w$,

And axiom (b) is equivalent to: $(0 < x < y \ \&0 < z < w) \Rightarrow xz < yw$.

Also, axiom (c) means the number can not be both greater than and less than another number. Axiom (d) means for any 2-different numbers one must be the largest.

3. The least upper bound axiom: A real number U is called upper bound of a set S of real numbers if for all $x \in S$ we have $x \leq U$. If an upper bound u can be found such that for all upper bounds U we have $u \leq U$, then u is called the least upper bound or supremum of S, abbreviated by $l.u.b.S$ or sup$S$.

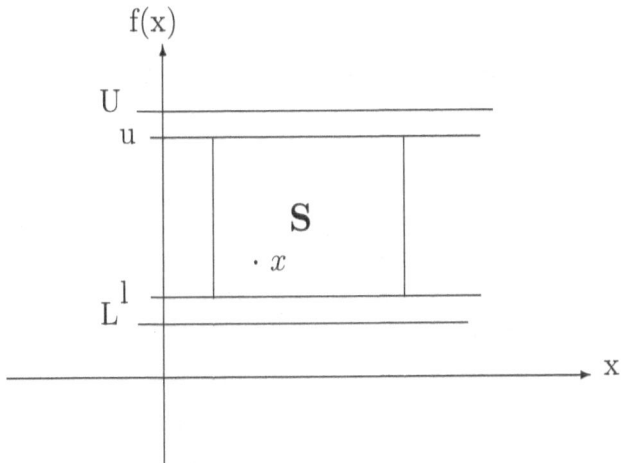

Theorem : If A is any non-empty subset of S that is bounded above, then A has a least upper bound in S, abbreviated by, l.u.b. or by sup$S$.

For example: if $A = \{1, 1.4, 1.41, 1.414, ...\}$, then (in R) the l.u.b for A is $\sqrt{2}$ which is not in the set..

4. The greatest lower bound axiom: A real number L is called a lower bound of a set S of real numbers if for all $x \in S$ we have $x \geq L$. If a lower bound l can be found such that for all lower bounds L we have $l \geq L$, then l is called the greatest lower bound abbreviated as $g.l.b.S$ or inf$S$.

For example: If $A = \{\frac{1}{2}, \frac{3}{4}, ., ., \frac{2^n-1}{2^n}, ...\}$, then the greatest lower bound for the set is $\frac{1}{2}$, and the lowest upper bound is 1.

Theorem : If A a non-empty set of real numbers, has a lower bound it has a greatest lower bound in S or g.l.b of S denoted by Inf$S$.

Proof: Let $B \subset S$ be the set of all $x \in S$ such that $-x \in A$ (i.e elements of B are the negatives of the elements of A. If U is a lower bound for A, then -U is an upper bound for B. For, if $x \in B$ then $-x \in A$ and so $U \le -x, x \le -U$.

Hence B is bounded above so that B has a l.u.b. If L is the l.u.b. for B then $-L$ is the g.l.b. for A.

Corollary:

If $R = L \cup U$ and if for each l in L and each u in U we have $l < u$, then either L has a greatest element(denoted by inf.S) or U has a least element (denoted by sup.S).

5. Archimedes axiom: For any real number a there is an integer n such that $a < n$.

Proof: Consider the set S of integers k such that $k \le a$. Since S has the upper bound a, then by axiom (3) it has a least upper bound u. Since u is the least upper bound for S, then $u - 1/2$ can not be an upper bound for S, and so there is a $k \in S$ such that $k > u - 1/2$. But $k + 1 > u + 1/2 > u$, and so $(k + 1) \notin S$. Since $k + 1$ is an integer not in S, we must have $k + 1$ greater than a by the definition of S.

Corollary :

Between any two real numbers is a rational ; that is , if $a < b$, then there is a rational r with $a < r < b$.

Proof: Suppose $a \le 0$, then by Archimedes axiom there is an integer $q > (b - a)^{-1}$ or $1/q < b - a$. The set of integers n such that $b < (n/q)$ is a nonempty set of positive integers, so it has a least element p , then

$$(p-1)/q < b < (p/q), \qquad (1.7)$$

and

$$a = b - (b-a) < (p/q) - (1/q) = (p-1)/q. \qquad (1.8)$$

Thus, $r = (p-1)/q$ lies between a and b. If $a < 0$, we can find an integer n such that $n > -a$. Then $n + a > 0$, and there is a rational r with $n + a < r < n + b$, and $r - n$ is a rational between a and b.

6. The completeness axiom: This axiom distinguish the real numbers from other ordered field. Every nonempty set S of real numbers which has an upper bound has a least upper bound. As a consequence to this axiom we have the following proposition:

In calculus, the proofs of many basic theorems such as : the existence of maximum and minimum, the intermediate value theorem, Rolles theorem, the mean value theorem, .. , etc, depends strongly on the so-called completeness property of the real numbers. To see this we will state these theorems, and give an example.

Existence of maximum and minimum: A function f(x) that is continuous on the closed interval $[a, b]$ has absolute extrema at the interval either at the endpoint of the interval (a, b) or at c where, $a < c < b$.

Example: Find the extreme maxima and minima of the function : $f(x) = 4x^3 - 6x + 7$, on the closed interval $I = [-1, 0]$.

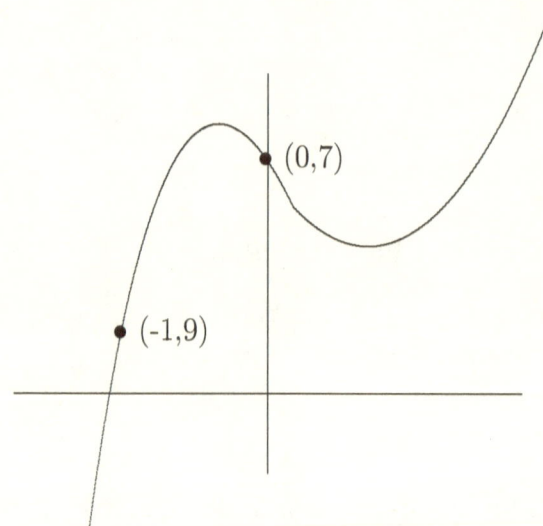

Solution: the first derivative of $f(x)$ is: $f'(x) = 12x^2 - 6$.
So, the first order critical points occur when, $12x^2 - 6 = 0$. Or at $x = \pm\frac{1}{\sqrt{2}}$. From this we see that $x = \frac{-1}{\sqrt{2}}$ is the only one that falls inside the interval $I = [-1, 0]$.
Looking for maximum and minimum we compute $f(x)$:
At the critical point: $x = \frac{-1}{\sqrt{2}} \rightarrow f(x) = 9.83$.
At the interval end point : $x = -1 \rightarrow f(x) = 9$.
And at the origin: $x = 0 \rightarrow f(x) = 7$.
From these information we see that:
The absolute maximum of $f(x)$ On the interval $I = [-1, 0]$ is $F(-\frac{1}{\sqrt{2}}) = 9.83$.
And the absolute minimum of f on the interval $I = [-1, 0]$ is $f(0) = 7$.

Intermediate Value Theorem:
Let f be continuous on the closed interval $I = [a, b]$ with $f(a) \neq f(b)$. If c is any number between $f(a)$ and $f(b)$, then there is at least one number x_0 in the interval (a, b) such that

$f(x_0) = c.$

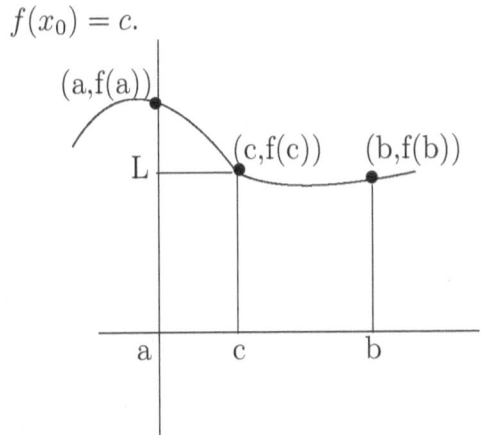

Example: Show that the equation : $x^3 - 2x^2 + x - 5 = 0$, has a solution between $x = 2$ and $x = 3$.

Solution: Let $f(x) = 2x^3 - x^2 + 4x - 6$. Since $f(x)$ is a polynomial, then continuous for all x.
Substituting x-values in $f(x)$ gives:
For $x = 2 \rightarrow f(2) = -3$, and for $x = 3 \rightarrow f(3) = 7$.
Then the Initial Value Theorem tells us that f assumes all the values between -3, and 7 as x ranges from 2,to 3. Since 0 is between $-3, 7$, it follows that there is a number x_0 in the open interval $(2, 3)$ such that: $f(x_0) = 0 \Rightarrow x_0$ is the root of the equation : $x^3 - 2x^2 + x - 5 = 0$ that lies between $x = 2$, and $x = 3$.

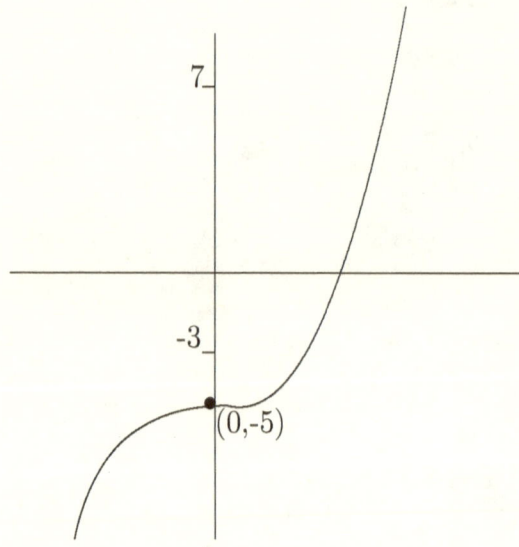

Rolles' Theorem: Suppose f is continuous at any point on $[a, b]$, and differentiable at any point of (a,b), with $f(a) = 0$, and $f(b) = 0$. Then there exist a number c in open interval (a, b) such that $f'(c) = 0$.

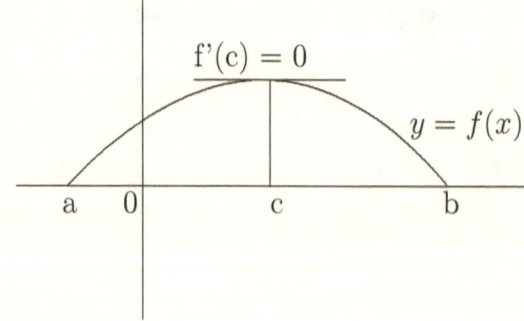

Example: For the given function:

$$f(x) = 1 - x - (3x - 1)^{1/3}.$$

on the interval $[0, 3]$, prove Rolle's Theorem.
solution: On the given interval: $f(0) = f(3) = 0$.

$$f'(c) = -1 + 1/3(3c - 1)^{-2/3}(3).$$

Solving for c gives:

$$1 = (3c - 1)^{-2/3}. \ then, \ c = 2/3.$$

Thus Rolle's conditions are satisfied.

Mean Value Theorem: Let f be continuous on the interval $[a, b]$ and differentiable on the interval (a, b). Then there exist a point c in the open interval (a, b) such that:

$$f'(c) = \frac{f(a) - f(b)}{b - a}.$$

Example: For the function :

$$f(x) = x^3 - 2x^2 + 3.$$

Find a number c in the interval $I = [-1, 4]$, that satisfy the Mean Value Theorem.
Solution: From the theorem we get: $f(4) = 35$, $f(-1) = 0$, and $f'(x) = 35/5 = 7$.

$$Then, \ f'(c) = 3c^2 - 4c = 7.$$
$$Or, 3c^2 - 4c = 7.$$
$$Solving \ for \ c \ gives \ c = \frac{4 \pm 10}{6}.$$
$$Or, c = -1, \ and \ c = 7/3.$$

But on the open interval $(-1, 4)$, $c = 7/3$ is the only one that satisfies the mean value theorem.

Now, after going through all these theorem , we can talk about the completeness axiom(property) or the least upper bound.

1.3.1   Inequalities

<u>Definition</u>: Let a and b be real numbers, and p positive numbers, then:

i) a is less than b , written as $a < b$, means $b - a$ is a member of p.

ii) b is greater than a, written as $b > a$, means $a < b$.

iii) $a \leq b$ means, $a < b$ or $a = b$.

iv) $b \geq a$ means, $b > a$ or $a = b$.

<u>Examples:</u>

1. For each real number (a) exactly one of the following is true:

i) $a > 0$.

ii) $a = 0$.

iii) $a < 0$.

From the definition above: $a > 0$ means $a - 0 = a \in P$, and $a < 0$ means $0 - a = -a \in P$. Hence exactly one of the statements: $a > 0, a = 0, a < 0$ is true.

2. If $x > 1$ then $x - 1$ is positive. In order to conclude that $x + 3 > 4$ we must show that: $(x + 3) - 4$ is positive.

But $(x + 3) - 4 = x - 1$. Hence if $x > 1$ then $(x + 3) - 4$ is positive , so $x + 3 > 4$.

3. For every real number p, and q:

$$p^2 + q^2 \geq 2pq. \tag{1.9}$$

We want to conclude that $(p^2 + q^2) - 2pq$ is ≥ 0.

Notice that: $(p^2 + q^2) - 2pq = p^2 - 2pq - q^2 = (p - q)^2$
But $(p - q)^2$ is the square of real numbers, which is positive or zero. Hence $p^2 + q^2 \geq 2pq$ is true.

4. If $a < b$, then the average, or the arithmetic mean of a, and b is between a, and b. That is, if $a < b$, then :

$$a < \frac{a+b}{2} < b. \qquad (1.10)$$

Proof:
From

$$a < \frac{a+b}{2}, \ we \ get: \ \frac{a+b}{2} - a = \frac{b-a}{2} \qquad (1.11)$$

Which is the product of two positive numbers $\frac{1}{2}$, and $(b - a)$, since $a < b$, then $(\frac{a+b}{2} - a)$ is positive.
Similarly,

$$\frac{a+b}{2} < b, then \ b - (\frac{a+b}{2}) \Rightarrow \frac{b-a}{2}. \qquad (1.12)$$

Then,

$$\frac{a+b}{2} < b \ is \ positive \ too. \qquad (1.13)$$

5. Suppose a, b, and c are real numbers, then the following conditions hold:
i) If $a < b$, and $b < c \Rightarrow a < c$.
ii) $c > 0$, $a < b$, IFF $ac < bc$.
iii) $a < b$, IFF $a + c < b + c$.
iv) $c < 0$, $a < b$, IFF $ac > bc$.

Proof:
i) Since $a < b$, and $b < c$ by definition means both $(b-a)$, and $(c-b)$ are both positive. Then using this fact it follows that $(c-a)$ is also positive:
$c - a = (c-a) + (b-a)$ which is positive. Then $a < c$.
(ii), (iii), and (iv) are left for the student to proof in the same manner as done in (i).

Corollary: Let a, b, and c be real numbers, then:
i) If $a < b$, and $c < d \Rightarrow a + b < b + d$.
ii) If $0 < a < b$, $0 < c < d \Rightarrow ac < bc$.

Proof:
i) From (iii) above , we get : $a + c < b + c$, and $b + c < b + d$. Then by (i) above gives: $a + c < b + d$.
ii) From $0 < a < b \Rightarrow a < b$.
$0 < c < d \Rightarrow c < d$. Then multiplying $a < b$ by the number c gives: $ac < bc$, and multiplying $c < d$ by b gives: $bc < bd$
Then using (i) gives: $ac < bd$.

6. Determine all the numbers x that satisfy the conditions:
$5x + 3 < 2$.
Solution:
If x satisfy this condition then by (iii)$5x < -1$, and by (iv) it implies: $x < -\frac{1}{5}$.
Now, we need to show that every number x that is less than $-\frac{1}{5}$ does satisfy the original condition. This means we will work backward as follows:
If $x < -\frac{1}{5}$, then applying condition (iv) gives: $5x < -1$, and by (iii) gives: $5x + 3 < 2$.
Thus, the solution is the set of all numbers less than $-\frac{1}{5}$, or in set solution notation: $\{x : 5x + 3 < 2\} = \{x : x < -\frac{1}{5}\}$.
This set represents the solution set of the inequality $5x + 3 < 2$.

7. Solve the inequality: $2x + 3 > x - 5$.
Solution:
By (iii) $2x+3 > x-5$ IFF $2x+3+(-1x-3) > x-5+(-1x-3)$
$\Rightarrow x > -8$. Thus the solution set is : $\{x : x > -8\}$.

8. Solve the inequality: $(x - 5)(x + 1) > 0$.
Solution: As was shown before:

$$(x - 5)(x + 1) > 0 \ IFF, \tag{1.14}$$

$x - 5 > 0$, and $x + 1 > 0$, or $x - 5 < 0$, and $x + 1 < 0$,
then $x > 5$ and $x > -1$, or $x < 5$, and $x < -1$.
Thus, $x > 5$, and $x > -1$ IFF $x > 5$.
and $x < 5$, and $x < -1$ IFF $x < -1$.
Hence the solution set is : $\{x : x > 5 \ or \ x < -1\}$

9. Solve the inequality: $x^2 - 6 > x$.
Solution is left for the student to prove.

Theorem:
Suppose a, and b are real numbers, and n is a positive integer.
If $0 < a < b \Rightarrow a^n < b^n$.
Proof is left for the student.

1.3.2 Absolute Values Inequality:

Definition: From the additive inverse property:
$a + (-a) = (-a) + a = 0$ for all real numbers. So the additive inverses are numbers that are the same distance from zero (origin) on the number line. This idea can also be expressed by saying that a number and its additive inverse have the same absolute value, this is symbolized by | |. Then we can express the absolute value as:

$$| a | = \begin{cases} a \ if \ a \geq 0 \\ -a \ if \ a < 0. \end{cases}$$

Based on this definition then the difference between two real numbers x, and y in R can be written as $\mid x - y \mid$ which represents the distance between x and y ,denoted as d, and is also known as the **Metric**, i.e $d(x,y) = \mid x - y \mid$.

The Metric space, also, is defined as the distance $d(x,y)$ between the elements $\{x, y, z\}$ of the space that satisfies the following conditions:

$$1. d(x,y) \;=\; 0. \; \textit{non negative conditions.}$$
$$2. d(x,y) \;=\; d(y,z). \; \textit{symmetric conditions.}$$
$$3. d(x,y) \;=\; 0, \; IFF \; x = y.$$
$$4. d(x,y) \;\leq\; d(x,y) + d(y,z), \; \textit{Triangle inequality.}$$

Note that the metric space was generalized from the Euclidean space where the distance between ordered m-tuple: $\{x_1, x_2, x_3, ., ., ., x_m\}$, and $\{y_1, y_2, y_3, ., ., y_m\}$ is defined as:

$$d(x,y) = \sqrt{(x_1 - y_1)^2 + (x_2 - y_2)^2 + ... + (x_m - y_m)^2}.$$

If we let $m = 1$, then the distance become: $d(x,y) = |x - y|$. And this is the absolute value $|a|$ with the conditions as mentioned above.

Theorem-1:
i) If $a \in R \Rightarrow \mid a \mid = \sqrt{a^2}$
ii) If $a \in R$, and $b \in R \Rightarrow \mid ab \mid = \mid a \mid \mid b \mid$

Proof:
i) If $a \in R$ this means $\mid a \mid = \sqrt{a^2}$.
Since $\sqrt{a^2} = |a|$ if $a \geq 0$, and $\sqrt{a^2} = |-a| = a$ if $a < 0$.
Then $\mid a \mid = \sqrt{a^2}$

ii) In the same way we can proof that:

$$| \, ab \,| = \sqrt{(ab)^2} = \sqrt{a^2}\sqrt{b^2} = |\, a \,| \, |\, b \,|$$

Theorem-2:
If $a \in R$, and $b \in R \Rightarrow |\, a + b \,| \leq |\, a \,| + |\, b \,|.$

Proof:
Since both $|\, a + b \,| \geq 0$, and $|\, a \,| + |\, b \,| \geq 0$.
Then: $|\, a + b \,| \leq |\, a \,| + |\, b \,|$ IFF $|\, a + b \,|^2 \leq (|\, a \,| + |\, b \,|)^2$

By the theorem above:

$$
\begin{aligned}
(|\, a \,| + |\, b \,|)^2 - |\, a + b \,|^2 &= |\, a \,|^2 + 2 |\, a \,| \, |\, b \,| + |\, b \,|^2 - (a+b)^2. \\
&= a^2 + 2 |\, ab \,| + b^2 - (a^2 + 2ab + b^2) \\
&= 2 |\, ab \,| - 2ab \\
&= 2(|\, ab \,| - ab).
\end{aligned}
$$

But $ab \leq |\, ab \,| \Rightarrow$ Hence, $(|\, a \,| + |\, b \,|)^2 - |\, a + b \,|^2 \leq 0$

So $|\, a + b \,|^2 \leq (|\, a \,| + |\, b \,|)^2$

Or $|\, a + b \,| \leq |\, a \,| + |\, b \,|$.

Examples:

1. Solve $|\, y + 1 \,| = y$ for y.

Solution: If $|\, y + 1 \,| = y \Rightarrow |\, y + 1 \,|^2 = (y+1)^2 = y^2$
Or $y^2 + 2y + 1 = y^2 \Rightarrow 2y + 1 = 0$
Then, $y = -\frac{1}{2}$ is the solution.
But if $y = -\frac{1}{2} \Rightarrow |\, y + 2 \,| = |\, -\frac{1}{2} + 2 \,| = |\, \frac{3}{2} \,| = \frac{3}{2}$ Which is $\neq y$.

2. Solve $|\, x - 4 \,| \leq |\, 3x + 1 \,|$ for x.

Solution: Since both sides are positive no matter what the

value of x is, then: $|x - 4|^2 \leq |3x + 1|^2$.
Which is equivalent to :

$$
\begin{aligned}
(x - 4)^2 &\leq (3x + 1)^2 \\
\Rightarrow (3x + 1)^2 - (x - 4)^2 &\geq 0 \\
\Rightarrow [(3x + 1) - (x - 4)][(3x + 1) + (x - 4)] &\geq 0 \\
\Rightarrow (2x + 5)(4x - 3) &\geq 0.
\end{aligned}
$$

Hence x is a solution   IFF $x \leq -\frac{5}{2}$ or $x \geq \frac{3}{4}$.

$$
\begin{array}{ccc}
 & -5/2 & 3/4 \\
+ & & + \\
\hline
 & - & \\
\end{array}
$$

Properties of Absolute Values: The student will remember these from Algebra classes:

$$
\begin{aligned}
&1.\ |x| \geq 0 \ , |x| = 0 \ IFF \ x > 0. &(1.15) \\
&2.\ |xy| = |x| \cdot |y| . &(1.16) \\
&3.\ \left|\frac{x}{y}\right| = \frac{|x|}{|y|}. &(1.17)
\end{aligned}
$$

3. If $\epsilon > 0$, then,

$$
\begin{aligned}
i)\ |x| &< \epsilon \ means \ -\epsilon < x < \epsilon &(1.18) \\
ii)\ |x| &\leq \epsilon \ means - \epsilon \leq x \leq \epsilon. &(1.19)
\end{aligned}
$$

4. The Triangle inequality : $|x + y| \leq |x| + |y|$ holds.

5. $|-x| = |x|; \ |x - y| = |y - x|.$

6. $|x|^2 = x^2; \ |x| = \sqrt{x^2}.$

7. $|x - y| \leq |x| + |y|.$

8. $|| x | - | y || \leq | x - y |$.

Theorem-1 : (The Triangle Theorem)

If a, and b are any two real numbers, then :

1. $| a + b | \leq | a | + | b |$

2. $| a - b | \geq || a | - | b ||$

3. $| a + b | \geq || a | - | b ||$

Proof:

1. To proof (1) use the following possibilities:

a) If $a \geq 0$, and $b \geq 0 \Longrightarrow a + b \geq 0$, so $| a + b | = a + b = | a | + | b |$.

b) If $a \leq 0$, and $b \leq 0 \Longrightarrow a + b \leq 0$, so $| a + b | = -a + (-b) = | a | + | b |$.

c) If $a \geq 0$, and $b \leq 0 \Longrightarrow a + b = | a | - | b |$.

d) If $a \leq 0$, and $b \geq 0 \Longrightarrow a + b = - | a | + | b |$.

Therefore, (1) holds in either case since:

$$| a + b | = \begin{cases} | a | - | b | & if \ | a | \geq | b | \\ | b | - | a | & if \ | b | \geq | a | \end{cases}$$

2. To proof (2) let : $a = a - b$ and substitute in :

$$| a + b | \ \leq \ | a | + | b | \ gives \qquad\qquad (1.20)$$

$$|a| \leq |a-b| + |b| \qquad (1.21)$$

Subtracting $|a-b|$ from both sides gives:

$$|a| - |a-b| \leq |a-b| - |a-b| + |b| \qquad (1.22)$$
$$|a| - |a-b| \leq |b| \qquad (1.23)$$
$$-|a-b| \leq |b| - |a|, or \qquad (1.24)$$
$$|a-b| \geq |a| - |b|. \qquad (1.25)$$

3. To proof (3), from the proof of (2) or (1.25) interchanging a and b yields,

$$|b-a| \geq |b| - |a| \qquad (1.26)$$
$$|-(a-b)| \geq |b| - |a| \qquad (1.27)$$
$$|a-b| \geq |b| - |a| \qquad (1.28)$$

Since $||a| - |b|| = |a| - |b|$ if $|a| > |b|$, then $|a-b| \geq ||a| - |b||$.

1.4 Exercise - 1

In each of the following sets find their members explicitly:

1. $\{x/x$ is a real number, and $2x - 3 = -4(x + 1) = 0\}$.

2. $\{n/n$ is an integer, and $3(4n - 1) = 4n + 2\}$.

3. $\{x/x$ is a real number, and $2x^2 - 2x - 3 = 1\}$.

4. $\{x/x$ is a real number, and $4x^2 - 6x + 1 = 0\}$.

5. State what each of the following means in terms of the difference of two numbers (all letters denote real numbers):
a) $a < b$.
b) $x > 3$.
c) $x + y \geq z - w$.
d) $x^2 - 3 < \pi/2$
e) $x^2 - 3 < -\frac{\pi}{2}$.
f) $-\frac{\pi}{2} < x + \sqrt{3} < \frac{\pi}{2}$.
g) $-\frac{\pi}{2} < x^2 - 3 < \frac{\pi}{2}$.
h) $\sqrt{xy} \leq \frac{x+y}{2}$.

6. Justify each of the following:
a) $5 < 8$.
b) $-5 \leq 1$.
c) $5 \geq 5$.

7. Prove or disprove each of the following statements:
a) $3 > 4$.
b) $2 < 2$.

8. If A is a real number, show that $A + 3 < A + 4$.
Does the result hold if $<$ is replaced with $>$?

9. Show that if $x > 0$ and $y < 0$, then $xy < 0$.

10. Solve each of the following inequalities:
a) $4x - 1 < 7$.
b) $x + 2 < x + 3$.
c) $\frac{2}{x} > 1$.

11. Solve each of the following inequalities:
a) $x^2 + x + 6 > 0$.
b) $(x + 2)(x - 5) > 0$.
c) $(2x - 5)(x - 1) < 0$.
d) $\frac{5-3x}{x-7} < 0$.
e) $\frac{x+7}{x-4} < 1$.

12. Solve each absolute value equation:
a) $\mid y + 2 \mid = 2x$.
b) $\mid x - 5 \mid = x$.
c) $\mid 5x + 1 \mid = \mid 3x - 7 \mid$.

13. Prove or disprove : If $a \in R$, $B \in R$, then $\mid A - B \mid \leq \mid A + B \mid$.

14. Solve each of the following inequalities, and graph the solution set.
a) $\mid 2 - u \mid > u$.
b) $\mid 7x - 5 \mid > 3 \mid x \mid$.
c) $\mid z + 5 \mid > z + 5$.

15. Prove the general triangle inequality given as,

$$\mid x_1 + x_2 + ... + x_n \mid \leq \mid x_1 \mid + \mid x_2 \mid + ... + \mid x_n \mid . \qquad (1.29)$$

16. Find the value of x that satisfies the following inequalities, and express your answer without the absolute values:
a) $\mid x - 5 \mid > x - 3$.

b) $|x + 7| < 4|x|$.

c) $|x - 9| < 7 < |7x - 27|$.

17. Sketch the graph of the following equations:
a) $y = x^2 . |x|$

b) $|y| < |y|$.

c) $|x| + |y| < 1.$

1.4.1   Rational Numbers

When we studied Algebra, we treated rational numbers as fractions denoted as $\frac{p}{q}$, where p, and q both integers, and $q \neq 0$. Given an integer $p > 1$, we can associate with each fraction at least one sequence: $0.a_1, a_2, a_3, ..., a_k$, where $0 \leq a_k \leq p-1$. for example, for $p = 1$ the fraction $\frac{1}{6} = 0.1666666667$, which present a fraction $\frac{m}{n}$ with $0 < m < n$. But this is just a decimal expression, then this means there is an association between fractions and decimals, and for this reason **R** is regarded as a metric space with distance d. To say that $\frac{m}{n}$ has a decimal expression means that attached to it there is a certain infinite series represented as:

$$S = \sum_{j=1}^{\infty} \frac{a_j}{10^j}, \tag{1.30}$$

where $0 \leq a_j \leq 9$, and a_j an integer.

1.4.2   Irrational Numbers

In the beginning of the chapter we gave $\sqrt{2}$ as example for irrational numbers, here we will give three important examples for irrational numbers:

a.   The remarkable irrational number called *phi* or more commonly " Golden ratio", denoted as ϕ which can be described as the solution to the equation:

$$\frac{A+1}{A} = \frac{A}{1}.$$

And, the positive value obtained $: A = \dfrac{1+\sqrt{5}}{2}.$

Using the square root algorithm, this can be approximated as follows:

$$\phi = 1 + \cfrac{1}{1 + \cfrac{1}{1 + \cfrac{1}{1 + \cfrac{1}{1 + \cfrac{1}{1 + \frac{1}{1 + \dots}}}}}}$$

b. The number $pi(\pi)$: This can be approximated in many different ways, we have used the approximation of $\pi = \frac{22}{7}$, or,

$$\pi \;=\; 4 \times (1 - \frac{1}{3} + \frac{1}{5} - \frac{1}{7} + \frac{1}{9} - \frac{1}{13} + \frac{1}{15}..)$$

$$Or, \;\pi \;=\; 2 \times \frac{2^2}{1 \times 3} \times \frac{4^2}{3 \times 5} \times \frac{6^2}{5 \times 7} \times \frac{8^2}{7 \times 9} \times \frac{10^2}{9 \times 11} \times ...$$

c. e the compound interest and biological growth, which can be approximated as,

$$e \;=\; (1 + \frac{1}{n})^n$$

$$=\; 1 + \frac{1}{1} + \frac{1}{2} + \frac{1}{6} + \frac{1}{24} + \frac{1}{120} + \frac{1}{720} + \frac{1}{5040} + ..$$

$$=\; \frac{1}{1} + \frac{1}{2} + \frac{1}{2 \times 3} + \frac{1}{6 \times 4} + \frac{1}{24 \times 5} + \frac{1}{120 \times 6}$$

$$+\; \frac{1}{720 \times 7} + ..$$

Geometric Constructions of some Irrational points:

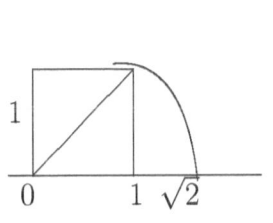

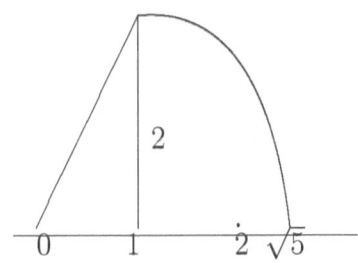

1.4.3   Cantor's Theory

George Cantor $(1845-1918)$ developed the theory of irrational numbers from the properties of regular sequences of irrational numbers.

Definition: The Cantor set **k** is the set of all numbers x in $[0, 1]$ which have a ternary expansion without the digit 1, this means the expansion of its sequence $\langle a_n \rangle$ is with $a_n \neq 1$. Thus the number $1/3 = 0.0222...$ and $2/3 = 0.20000...$ are in **k**, but any x such that $1/3 < x < 2/3$ is not in k.

Every number in the cantor set has a ternary expansion (in scale of 3) of the form,

$$f(x) = \sum_{n=1}^{\infty} \frac{b_n}{3^n}, \quad where\; b_n is\; either\, 0\; or 2. \tag{1.31}$$

Also, each number in $[0, 1]$ has a binary expansion (in scale of 2) of the form,

$$f(x) = \sum_{n=1}^{\infty} \frac{a_n}{2^n}, \quad where\; a_n is\; either\, 0\; or 1. \tag{1.32}$$

Where f maps the cantor ternary set into the interval $[0, 1]$.

The cantor set: The cantor set **k** has the following qualities:

1. **k** is nowhere dense.
2. **k** is perfect.
3. **k** has measure zero.
4. **k** has cardinality N.
5. **k** is closed set.

Construction of Cantor's Set:

Start with the closed unit interval $[0,1]$ Trisect the interval at points $1/3, 2/3$.

$$\vdash\!\dashv$$
$$0 \quad 1/3 \quad 2/3 \quad 1$$

The sets are: $\underbrace{0, 1/3}$ $\underbrace{1/3, 2/3}$ $\underbrace{2/3, 1}$.

Then remove the middle interval $(1/3, 2/3)$, and keep $\underbrace{0, 1/3}$, $\underbrace{2/3, 1}$, and let:

$$K_1 = [0, 1/3] \cup [2/3, 1]. \tag{1.33}$$

Now, trisect each subsection again:

$$0 \quad 1/9 \; 2/9 \; 3/9 \; 4/9 \; 5/9 \; 6/9 \; 7/9 \; 8/9 \; 1$$

The sets are:

$\underbrace{0, 1/9}$ $\underbrace{1/9, 2/9}$ $\underbrace{2/9, 3/9}$ $\underbrace{3/9, 4/9}$ $\underbrace{4/9, 5/9}$ $\underbrace{5/9, 6/9}$
$\underbrace{6/9, 7/9}$ $\underbrace{7/9, 8/9}$ $\underbrace{8/9, 1}$.

Removing the open middle third,

$\underbrace{0, 1/9}, \underbrace{2/9, 3/9}, \underbrace{4/9, 5/9}, \underbrace{6/9, 7/9}, \underbrace{8/9, 1}$ we obtain,

$$K_2 = [0, 1/9] \cup [2/9, 3/9] \cup [6/9, 7/9] \cup [8/9, 1]. \tag{1.34}$$

Continuing in the same manner we obtain the sets $K_1, K_2, ...$ The Cantor set denoted by **k** is the intersection of $K_1, K_2, ..$ When iterated, these steps lead to a nested sequence of closed sets $K_n : K_1 \supset K_2 \supset K_3 \supset ...$

Each K_n consisting of 2^n closed intervals of length 3^{-n}.
Then the set

$$K = \bigcap_{n \in \aleph} K_n. \tag{1.35}$$

Is the Cantor set.
To show that **k** is nowhere dense in $[0, 1]$ we see that no interval
of the form:

$$\left(\frac{2n-1}{3^m}, \frac{2n}{3^m}\right) \quad (m, n \in \aleph), \tag{1.36}$$

intersects **k**. Also, given an arbitrary interval $(a, b) \subset [0, 1]$,
then the inequalities,

$$a < \frac{2n-1}{3^m} < \frac{2n}{3^m} < b, \tag{1.37}$$

Multiplying (1.37) by 3^m and adding 1 gives:

$$3^m a < 2n - 1 < 2n < b\,3^m.$$
$$3^m a + 1 < 2n < b\,3^m + 1.$$

Then (1.37) is satisfied when $3^{-m} < b - a$, and
$3^m a + 1 < 2n < 3^m b + 1$.

Selecting a sufficiently large value of m to assure that the interval $(3^m a - 1, 3^m b)$ contains an even integer, we find every open
interval in $[0, 1]$ contains an interval which does not intersect
k. This proves (1) listed above.
To prove (2), we begin by the observation that k is closed: this
follows from the following theorem.

Theorem-1: For A any number, B closed, if $A \subset B$, then
$\overline{A} \subset B$.

Proof: The set will be shown to be perfect if it is shown that it has no isolated points as follows:

Let k be an arbitrary point of k; let $N(k, r)$ be any neighborhood of k. Each set k_n contains a component I_n such that

$$k = \bigcap_{n \in \eta} I_n. \tag{1.38}$$

Because, $m(I_n) = 3^{-n}$, it follows that $I_n \subset N(k; r)$ whenever $3^{-n} < r$. But each I_n contains at least two members of k, then (2) is established.

Another way to proof quality (2) of cantor's Theorem, is to proof that k is perfect, as shown in the following theorem.

Theorem-2: A perfect set is either empty set or uncountable.

Proof: Suppose k is countable, and has members $k_1, k_2, ...,$ and assume at the moment there is a nested sequence of compact intervals, I_n,

$I_1 \supset I_2 \supset I_3 \supset ...$, such that for each value of n, $k_n \notin I_n$. But $J_n = I_n \cap k \neq \emptyset$, then no point of k is included in every set of J_n and hence,

$\bigcap_{n \in \eta} J_n = \emptyset$.

To prove(3) from the cantor set qualities, Let the intervals be I_n then the measure of $E \backslash k$ (the countable union of disjoint open intervals)is defined to be ,

$$\mu = \sum_{n=1}^{\infty} m(I_n) \tag{1.39}$$

Note: E represents the closed set $[0,1]$.

The measure of k in the first step is taken to be $1 - \mu$. In obtaining K_1 we remove an interval of length $1/3$; in the next step we remove two intervals, each of length 3^{-2}.

In general , 2^{n-1} intervals of length 3^{-n} each have been re-
moved in the nth step. Thus, the total length of the intervals
eliminated in this step is:

$$3^{-1} + 2.3^{-2} + ... + 2^{n-1}.3^{-n} = 1 - (2/3)^n, \qquad (1.40)$$

and (3) is proved. To prove (4) we consider the number $x \in$
$[0, 1]$ to the base 3 represented as,

$$x = \sum_{n=1}^{\infty} a_n 3^{-n} \ (a_n = 0, 1, 2). \qquad (1.41)$$

Apparently if $x \in E\backslash k_1$, $(E = [0, 1])$ then it is necessary that
$a_1 = 1$.

Setting:
$1/3 = 0.0222...$, and
$2/3 = 0.2000...$,
it follows that for all points in $x \in K_1$: $a_1 = 0, 2$. In a same
way, we realize that if $x \in E\backslash k_2$, then $a_2 = 1$ and with suitable
normalization we find $a_2 = 0, 2$ for all points $x \in K_2$. Continuing in this manner, we can conclude the following:
$x \in K$ Iff the entries a_n in its expansion to the base 3 assume
the value 0, and 2; that is, each such $x \in K$ has a representation $x = 0.a_1 a_2 a_3...$, where $a_n = 0, 2$.
Using base 2, each $x \in E$ has a representation $x = 0.b_1 b_2 b_3....$
in which $b_n = 0, 1$. The association $0 \longleftrightarrow 0$, $2 \longleftrightarrow 1$ establishes a one-to-one correspondence between k and E, and thus
k is uncountable.

1.5 Sets

Sets are used to group objects together, and this helps to study a collection of elements in an organized fashion.

1.5.1 Set Notation

First we begin with the notation of sets. If A is a set and x is an object in the set A, then the statement "x belongs to A" can be abbreviated as $x \in R$, while the statement "x does not belong to A" can be abbreviated as $x \notin R$. The symbol that is used for set is the braces { }. For example, the set whose members are 1, 2, and 3 is denoted by $\{1, 2, 3\}$. The members of the set can also be written as: $\{2, 3, 1\}$ or $\{3, 2, 1\}$ without affecting the set size.

The symbol $\forall x \in R \Rightarrow$ *means for every real number x.*
The symbol $\forall x \in Z \Rightarrow$ *means for every integer x.*
The symbol $\exists x \in Z \Rightarrow$ *means there exists an integer x.*

Sets are written in two type of notations:

1. Roster notation such as: $A = \{a, b, c\}$, or
2. Builder notation such as: $A = \{x \mid x \ even \ number \ < 9\}$

1.5.2 Description of sets

Empty Set:
A set with no elements inside is called "empty set", and is denoted by "ϕ" or { }, and its cardinal number is $n(s) = 0$.

Subset:
For two sets A, and B: Set A is called to be subset of set B, if and only if every element in A is also element in B.
The symbol for subset is $\Rightarrow \subseteq$. The above statements can be

described symbolically as:

$A \subseteq B$ and is read as: *A is a subset of B.*

$A \subseteq B$ IFF $\forall x \{x \in A \rightarrow x \in B\}$.

Proper Subset:

For two sets A, and B: Set A is called to be proper subset of set B, if and only if every element in A is also element in B, but not the other way around, because $A \neq B$.

The symbol for proper subset is $\Rightarrow \subset$. The above statements can be described symbolically as:

$A \subset B$ and is read as: *A is a proper subset of B.*

$A \subset B$ IFF $\forall x \{x \in A \rightarrow x \in B\} \wedge \exists x \{x \in B \wedge x \notin A\}$.

Rules of Subsets:

For every set A, the following rules apply:

1. Empty set is the subset of set $A \rightarrow \phi \subseteq A$.
2. Every set is a subset of itself $\rightarrow A \subseteq A$.

Proofs:

1. To proof $\rightarrow \phi \subseteq A$, means we have to show that :
every element in ϕ is also element in A, or Symbolically:
$\phi \subseteq A$ IFF $\forall e \{e \in \phi \rightarrow e \in A\}$.
Since empty set has no elements in it then:
$\{e \in \phi \rightarrow e \in A\} = \{F \rightarrow T\} = \text{T}$.

2. To proof $\rightarrow A \subseteq A$, means we have to show that :
every element in A is also element in A, or Symbolically:
$A \subseteq A$ IFF $\forall e \{e \in A \rightarrow e \in A\}$.
$\{e \in A \rightarrow e \in A\} = \{T \rightarrow T\} = \text{T}$.

The following table shows the truth values for the Logic Statement : $if, then$ or $A \to B$:

A	B	$A \to B$
T	T	$T \to T = T$
T	F	$T \to F = F$
F	T	$F \to T = T$
F	F	$F \to F = F$

We also show tables of the connectives AND ($\wedge$),and OR ($\vee$) that might be used in some problems.
The Conjunction table for(AND):

A	B	$A \wedge B$
T	T	$T \wedge T = T$
T	F	$T \wedge F = F$
F	T	$F \wedge T = F$
F	F	$F \wedge F = F$

The Disjunction Table for(OR):

A	B	$A \vee B$
T	T	$T \vee T = T$
T	F	$T \vee F = T$
F	T	$F \vee T = T$
F	F	$F \vee F = F$

Equal Sets: If two sets A , and B contains the same elements, with the same cardinal number, then the sets are said to be equal. To proof that set A = set B means to show that : $A \subseteq B$, and $B \subseteq A$. Then symbolically we can write:
$A = B \ IFF \ \forall x \{x \in A \to x \in B\} \wedge \forall x \{x \in B \to x \in A\}$. Or,
To show that: $A = B \ IFF \ \forall x \{x \in A \longleftrightarrow x \in B\}$.

Set size: The set size is measured by counting the number of elements in the set, and it is called the cardinal number or cardinality, and is symbolized by $n(s)$.

Example: If $A = \{a, b, c, d, e, f\}$ then $\rightarrow n(A) = 6$.

Expansion of sets:
To expand a set means to write all its subsets, as shown on the following example:

Example: Expand the set $\{a, b\}$ into its components.

Solution: The set $\{a, b\}$ has the following components:
$\{\ \}$ the empty set.
$\{a,\ \}$ one component a.
$\{\ , b\}$ one component b.
$\{a, b\}$ two components a, *and* b.

How to read or describe a symbolic statement::

The statement: $\forall x \in R\{x^2 \geq 0\}$ is read as: The square of every real number is positive $\rightarrow$ *it is a true statment..*

The statement : $\exists x \in Z\{x^2 = 1\}$ is read as: There exist an integer x whose square is 1. $\rightarrow$ *it is a True statment.*

The statement: $\{x/x$ *is an odd natural number* $\}$ is the set of all natural numbers. This can be described as: For some natural number n , $x = 2n - 1$.

The statement: $\{x/$ *for some integer* m $, x = m\pi\}$, can be described as: The set of all multiples of π.

The statement: $\{x/x \text{ is a real number }, \text{and } x^2 = 9\}$, can be described as: The set of all real numbers with square of 9.

1.5.3 Set Operations

Here we will consider member of sets that comes from a specific collection of elements, called the universal set, where the universal set is the real numbers.

1. Union:
The set of all elements (or points) which belong to either subset A or subset B or both A and B is called the union of A and B and is denoted by $A \cup B$.

2. Intersection:
The set of all elements which belong to both A and B is called the intersection of A and B and is denoted by $A \cap B$.

3. Difference:
The set consisting of all elements of A which do not belong to B is called the difference of A and B denoted by $A - B$.

4. Complement:
If $B \subset A$, then $A - B$ is called the complement of B_A relative to A and is denoted by $\overline{A}$. If $A = \mathcal{U}$, the universal set, we refer to $\mathcal{U} - B$ as simply the complement of B and denote it by $\overline{B}$. The complement of $A \cup B$ is denoted by $\overline{(A \cup B)}$

Some Theorems Involving Sets:

Theorem-1: $A \cup B = B \cup A$
Theorem-2: $A \cup (B \cup C) = (A \cup B) \cup C = A \cup B \cup C$
Theorem-3: $A \cap B = B \cap A$
Theorem-4: $A \cap (B \cap C) = (A \cap B) \cap C = A \cap B \cap C$

Theorem-5: $A \cap (B \cup C) = (A \cap B) \cup (A \cap C)$
Theorem-6: $A \cup (B \cap C) = (A \cup B) \cap (A \cup C)$
Theorem-7: $A - B = A \cap \overline{B}$
Theorem-8: If $A \subset B$, then $\overline{B} \subset \overline{A}$
Theorem-9: $A \cup \emptyset = A$, $A \cap \emptyset = \emptyset$
Theorem-10: $A \cup \mathcal{U} = \mathcal{U}$, $A \cap \mathcal{U} = A$
Theorem-11: $\overline{(A \cup B)} = \overline{A} \cap \overline{B}$, De Morgan's first law.
Theorem-12: $\overline{(A \cap B)} = \overline{A} \cup \overline{B}$, De Morgan's second law.
Theorem-13: $A = (A \cap B) \cup (A \cap \overline{B})$, for any set A and B.

1.5.4   Application of Set Operations

a. Union with symbol $\cup$.
b. Intersection with symbol $\cap$.

<u>Definition:</u>
a.   Union of two sets A ,and B is defined as the set that contains elements that are in A or in B or in both, and symbolically can be written as:
$A \cup B = \{x \mid x \in A \lor x \in B\}$.
This can be represented in Venn-Diagram as follows:

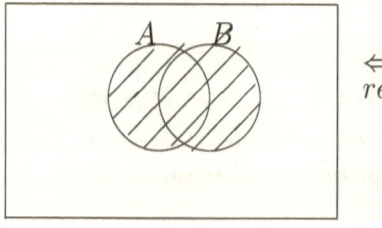

$\Longleftarrow$ *shaded area represents* $A \cup B$

Example: Find the union of the two sets: $\{a, b, c\}$, and $\{a, d, e\}$.

Solution:
The union for the two sets $= \{a, b, c\} \cup \{a, d, e\} = \{a, b, c, d, e\}$.

b. Intersection of two sets A ,and B is defined as the set that contains elements that are in both A and B, and symbolically can be written as:
$A \cap B = \{x \mid x \in A \wedge x \in B\}$ .This can be represented in Venn-Diagram as follows:

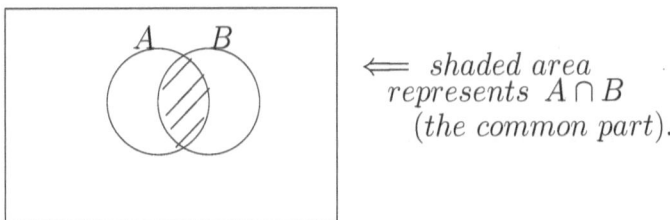

$\Longleftarrow$ *shaded area*
represents $A \cap B$
(the common part).

Example: Find the intersection of the two sets: $\{a, b, c\}$, and $\{a, d, e\}$.
Solution:
The intersection of the two sets $= \{a, b, c\} \cap \{a, d, e\} = \{a\}$.

Complement of the Set:

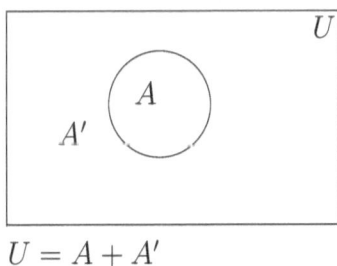

$U = A + A'$

A' *is complement of* $A = \{x \in U \mid x \notin A\}$, similarly,
A *is the complement of* $A' = \{x \in U \mid x \notin A'\}$.
Disjoint Sets:
Two sets that do not share any elements are called
" Disjoint":

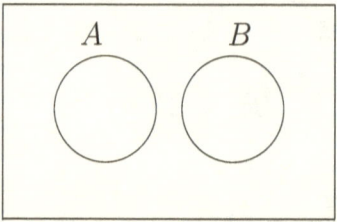

In the Venn-Diagram A, and B are disjoint. and
$A \cap B = empty\ set = \{\ \} = \phi$

Difference between two sets:

Definition: If A, and B are two sets, then : $A - B$ is the
set of all elements in A but not in B, or symbolically :
$A - B = \{x \mid x \in A \wedge x \notin B\}$.

Example: If $A = \{a, b, c\}, B = \{c, d, e\}$,
then : $A - B = \{a, b, c\} - \{c, d, e\} = \{a, b\}$
Or: $A - B = \{x \mid x \in A \wedge x \notin B\} = A \cap B'$.

De Morgan's Laws:

For sets A and B:
1. $(A \cup B)' = A' \cap B'$
2. $(A \cap B)' = A' \cup B'$,

Proofs: 1. Since,

$$\begin{aligned}
A \cup B &= \{x \mid x \in A \cup B\}. \\
Then, (A \cup B)' &= \{x \mid x \notin A \cup B\}. \\
&= \{x \mid \sim [x \in (A \cap B)]\}. \\
&= \{x \mid \sim (x \in A \cup x \in B)\}.
\end{aligned}$$

$$= \{x| \sim (x \in A) \cap \sim (x \in B)\}.$$
$$= \{x|x \notin A \cap x \notin B\}.$$
$$= \{x|x \in (A' \cap B')\}.$$
$$Thus, (A \cup B)\prime = A' \cap B'.$$

This can also be proofed using subset notations as follows:

$$If, \quad x \in (A \cup B)' \quad \rightarrow \quad x \notin (A \cup B).$$
$$\rightarrow \quad x \notin A \cap x \notin B.$$
$$\rightarrow \quad x \in A' \cap x \in B'.$$
$$\rightarrow \quad x \in A' \cap B'.$$
$$Then, \quad (A \cup B)' \subset A' \cap B'.$$

In the same manner the reverse of the identity can be proofed.

The Venn-Diagram for $(A \cup B)\prime$ is as shown below:

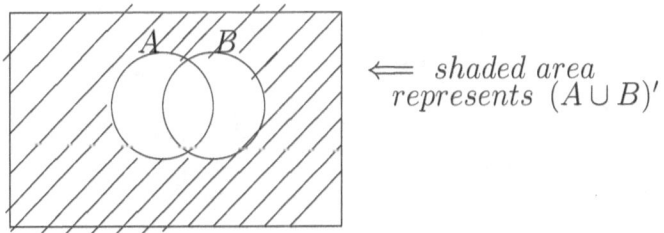

$\Longleftarrow$ *shaded area*
represents $(A \cup B)'$

2. Since,

$$A \cap B = \{x|x \in A \cap B\}.$$
$$Then, (A \cap B)\prime = \{x|x \notin A \cap B\}.$$
$$= \{x| \sim [x \in (A \cap B)]\}.$$
$$= \{x| \sim (x \in A \cap x \in B)\}.$$

$$= \{x| \sim (x \in A) \cup \sim (x \in B)\}.$$
$$= \{x|x \notin A \cup x \notin B\}.$$
$$= \{x|x \in (A' \cup B')\}.$$

Thus, $(A \cap B)' = A' \cup B'.$

The Venn-Diagram for $(A \cup B)'$ is as shown below:

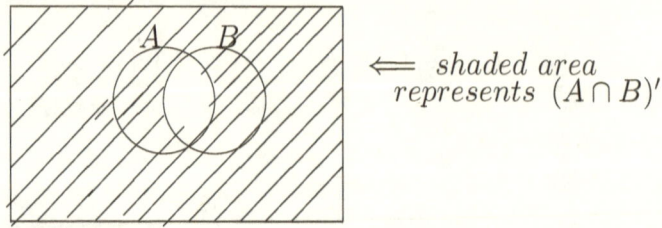

$\Longleftarrow$ *shaded area represents* $(A \cap B)'$

Example: using Venn-Diagram, Proof the following Identity:

$$A \cap (B \cup C) = (A \cap B) \cup (A \cap C). \qquad (1.42)$$

Solution: We will use algebraic Method to solve, to make it easier for the student to understand and relate it to the set notations: Using Venn- Diagram shown in the graph, we label the regions and state the sets first:

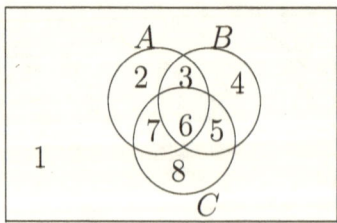

From the above Venn-Diagram, we can write the following sets:

$$U = \{1, 2, 3, 4, 5, 6, 7, 8\}.$$
$$A = \{2, 3, 6, 7\}.$$

$$B = \{3,4,5,6\}.$$
$$C = \{5,6,7,8\}.$$

We will start with the right-side of the Identity and prove it is equal to the left side:

$$R.S. \rightarrow (A \cap B) \cup (A \cap C) =$$
$$(\{2,3,6,7\} \cap \{3,4,5,6\}) \cup (\{2,3,6,7\} \cap \{5,6,7,8\}).$$
$$= \{3,6\} \cup \{6,7\}.$$
$$= \{3,6,7\}.$$
$$L.S. \rightarrow A \cap (B \cup C) = \{2,3,6,7\} \cap (\{3,4,5,6\} \cup \{5,6,7,8\}).$$
$$= \{2,3,6,7\} \cap \{3,4,5,6,7,8\}.$$
$$= \{3,6,7\}.$$
$$Then \ , L.S = R.S$$

Example: Proof the Identity of (1.42) using set notations:
Solution:: We still have to show that the right side is equal to the left side:

$$R.S. \rightarrow (A \cap B) \cup (A \cap C)$$
$$= \{x | x \in (A \cap B) \cup x \in (B \cap C)\}.$$
$$= \{x | x \in A \cap x \in B \cup x \in B \cap x \in C\}.$$
$$= \{x | x \in A \cap x \in B \cup C\}.$$
$$= \{x | x \subset (A \cap B \cup C)\}.$$
$$= A \cap B \cup C.$$
$$L.S. \rightarrow A \cap (B \cup C) = \{x | x \in A \cap (B \cup C)\}.$$
$$= \{x | (x \in A) \cap (x \in (B \cup C)\}.$$
$$= \{x | (x \in A) \cap (x \in B) \cup (x \in C)\}.$$
$$= \{x | (x \in A) \cap ((x \in B) \cup (x \in C))\}.$$
$$= \{x | x \in (A \cap (B \cup c)\}.$$
$$= A \cap B \cup C.$$
$$Hence, \ L.S = R.S.$$

Example:: Prove that for any set A and B we have:
$A = (A \cap B) \cup (A \cap \overline{B})$.

Solution:

$$
\begin{aligned}
A &= \{x : x \in A\} & (1.43) \\
&= \left\{x : x \in A \cap B \text{ or } x \in A \cap \overline{B}\right\} & (1.44) \\
&= (A \cap B) \cup (A \cap \overline{B}). & (1.45)
\end{aligned}
$$

Using the Venn-Diagram method , we can use the same sets for A, B, and C and the same steps as the previous example: The left-side of the statement is, A, equal to,

$$\{2, 3, 6, 7\}. \qquad (1.46)$$

The right-side of the statement is, $(A \cap B) \cup (A \cap \overline{B})$, equal to,

$$
\begin{aligned}
&= (\{2,3,6,7\} \cap \{3,4,5,6\}) \cup (\{2,3,6,7\} \cap \{1,2,7,8\}) \\
&= \{3,6\} \cup \{2,7\}, \\
&= \{2,3,6,7\}.
\end{aligned}
$$

Which is equal to the left side of the statement.

1.5.5 Summary of some theorems and Set operations

To better understand the concept of limits, continuity, differentiability, and integrabelity that was studied in calculus, it requires a better understanding of the real numbers, and the real line R.

1.5.6 Some Theorems on Point Sets

Theorem - 1: The complement of an open set is closed and the complement of a closed set is open.

Theorem - 2: The union of any number of open sets is open and the intersection of a finite number of open sets is open.

Theorem - 3: The union of a finite number of closed sets is closed and the intersection of any number of closed sets is closed.

Theorem - 4: Every open set on the real line can be expressed as a countable union of disjointed open intervals unique except as to the order of the intervals.

Theorem - 5: A set is called compact if every open covering of S has finite sub-covering. For R^n compact is equivalent to closed and bounded.

Examples:

1. *If* $A = \{x/x \text{ is a real number }, \text{and } x^2 = 9\}$, *and*

$B = \{3, -3\}$,

then, $A = B$ *or have exactly the same members.*

Proof: If $x \in A$, then x is a real number, and $x^2 = 9$. Solving $x^2 = 9$ gives:

$$x^2 - 9 = 0.$$
$$(x - 3)(x + 3) = 0.$$
$$\textit{Then, either } x = 3 \textit{ or } x = -3.$$

i.e $\{-3 \vee 3\} = \{-3, 3\}$ *all real numbers.*

Then every member of B is a member of A too.

2. Let $A = \{x/ (x - 2)(x - 1) = 0\}$, and let $B = \{1, 2\}$, then $A = B$.

Proof: As done in the previous example. If $x \in A$, then x is a real number, and $(x - 2)(x - 1) = 0$.
Solving $(x - 2)(x - 1) = 0$ gives:

$$(x - 2)(x - 1) = 0.$$
$$Then, \; either \; , x = 2 \; or \; x = 1.$$
$$i.e \; \{1 \vee 2\} = \{1, 2\} \; all \; real \; numbers.$$

Then every member of B is a member of A too or $\rightarrow A = B$

3. Is it true that $\{a\} = a$?

Proof: $\{a\}$ is a set with one element a or cardinal number equal 1. But a is a real constant, and these two are not equal to each other. Thus the statement is False.

4. Let,

$$A = \{x | x \; is \; a \; real \; number\}, and$$
$$B = \{x | \; 3x^2 - 8x + 4 = 0\}.$$

Show that $B \subset A$.

<u>Solution</u>:
Suppose $x \in A$ then by the quadratic formula:

$$x = \frac{8 \pm \sqrt{(-8)^2 - 4(4)(3)}}{6} = \frac{8 \pm 4}{6}. \qquad (1.47)$$

Therefore, $x = \frac{2}{3}$ or $x = 2$.

$$So, x = \{\frac{2}{3}, 2\} \in B.$$
$$Thus, A \; \subset \; B.$$

and since the elements $2/3$, and 2 are real numbers, then $x \in B$ thus every element of B belongs to A and so $B \subset A$. It follows that $A = B$.

5. If $A = \{x|x$ is a real number, and $1 - 2x = \sqrt{x}\ \}$. Find explicitly the members of A.

Solution:
Suppose $x \in A$,
then, $1 - 2x = \sqrt{x}$, squaring both sides and simplifying,
$(1 - 2x)^2 = (\sqrt{x})^2$, gives: $4x^2 - 5x + 1 = 0$.
Then, by quadratic formula, $x = \frac{1}{4}, or\ x = 1$.
Now, we know that 1, and $\frac{1}{4}$ are members of A, but we can not say $A = \{1, \frac{1}{4}\}$, we have to check first.
If $x = 1$, then $1 - 2x = -1 \neq 1$. So $1 \not\subset A$.
And if $x = \frac{1}{4}$, then $1 - 2x - \sqrt{x} = 0$. So $\frac{1}{4} \in A$, therefore, $A = \frac{1}{4}$, that is, A is the set whose members is $\frac{1}{4}$.

1.5.7 Countable and Non-Countable Sets

Corollary(1)[1]: The set R of all real numbers is not-countable.

Proof: Suppose the converse is true, that is R is countable. Then the set $R = \{x_1, x_2, ...\}$.
If we let the intervals,
$I_1 = \left\{x_1 - \frac{1}{2}, x_1 + \frac{1}{2}\right\}$,
$I_2 = \left\{x_2 - \frac{1}{4}, x_2 + \frac{1}{4}\right\}$, then in general for $n > 0$,
$I_n = \{x_n - 2^{-n-1}, x_n + 2^{-n-1}\}$. Then,

$$
\begin{aligned}
Length\ of\ I_n &= x_n + 2^{-n-1} - (x_n - 2^{-n-1}) \\
&= x_n - x_n + 2^{-n-1} + 2^{-n-1} = 2^{-n}.
\end{aligned}
$$

---

[1]Corollary is a proposition that follows with little or no proof required from already proven result.

So the sum of the lengths of all the I_n is :
$2^{-1} + 2^{-2} + 2^{-3} + ... = 1 \ (n \geq 1)$.   But $x_n \in I_n$ so that
$R = \bigcup_{n=1}^{\infty} \{x_n\} \subset \bigcup_{n=1}^{\infty} I_n$, but then the whole real line, whose
length is infinite, would be covered by union of intervals whose
length add up to 1, and this seems to be contradiction.

Corollary(2): The set of all rational numbers is countable.
Proof: Let the rational number E_n be:
$E_n = \{\pm N/n\} = \{0/n.1/n. - 1/n.2/n, -2/n.....\}$,
where N is a whole number such as $0, \pm 1, \pm 2, ...etc$, and $n \geq 1$.
Then the set of all rational numbers equal the union of,

$$
\begin{aligned}
E_n &= U_{n=1}^{\infty} E_n \\
&= \sum \{0/n, 1/n, -1/n.2/n, -2/n = 0.
\end{aligned}
$$

But 0 is a countable number, hence the set of all rationals is
the countable union of countable set.

Note To help understanding corollary-2 better, one can thing
about the (1-1) correspondence between the rational numbers
and the natural numbers presented in the following fashion:

$$
\begin{array}{cccccccccccc}
0 & 1 & \frac{1}{2} & \frac{1}{3} & \frac{2}{3} & \frac{1}{4} & \frac{3}{4} & \frac{1}{5} & \frac{2}{5} & \frac{3}{5} & \frac{4}{5} & \frac{1}{6} \Leftarrow \text{Rational numbers.} \\
\updownarrow & \updownarrow & \updownarrow & \updownarrow & \updownarrow & \updownarrow & \updownarrow & \updownarrow & \updownarrow & \updownarrow & \updownarrow & \updownarrow \\
1 & 2 & 3 & 4 & 5 & 6 & 7 & 8 & 9 & 10 & 11 & 12 \Leftarrow \text{Natural numbers.}
\end{array}
$$

This also proves that the set of rational numbers is countable
on the interval $[0, 1]$.

Example:
Show that the set of odd positive integers is a countable set.
Solution:
Consider the function $f(x) = 2x\text{-}1$ from S^+ to the set of odd
and positive integers. To show it is countable, we have to show

there is a one-to-one correspondence between these two sets. First we will show its one-to-one and onto:

If the set is one-to-one, suppose that $f(x) = f(y)$ then , $2x - 1 = 2y - 1$, which leads to $x = y$. To show its onto: Suppose that p is an odd positive integer, then p is less than an even integer $2k$, where k is a natural number. Hence $t = 2k - 1 = f(k)$. The one-to-one can be represented as follows:

1 2 3 4 5 6 7 8 9 . . $\Leftarrow$ as positive numbers.

$\updownarrow \updownarrow \updownarrow \updownarrow \updownarrow \updownarrow \updownarrow \updownarrow$

1 3 5 7 9 11 13 15 17 . . $\Leftarrow$ as odd positive numbers.

Example:

Show that the set of positive rational numbers is countable.

Solution:

Rational numbers are represented generally as a quotient $\frac{p}{q}$, where both p, and q are integers, and $q \neq 0$. Now, we will list all the positive rational numbers as sequences represented by : $r_1, r_2, r_3, .$, and will arrange them in columns based on p, and q, in the following way:

Column-1: $p = 1$, and $q = 1, 2, 3, ..$ with the sum of $2, 3, 4, ...$

Column-2: $p = 2$, and $q = 1, 2, 3, ..$ with the sum of $3, 4, 5, ..$

and so on as shown both separately:

2 3 4 5 6	$\frac{1}{1}$	$\frac{2}{1}$	$\frac{3}{1}$	$\frac{4}{1}$	$\frac{5}{1}$	$\frac{6}{1}$	$\frac{7}{1}$
3 4 5 6 7	$\frac{1}{2}$	$\frac{2}{2}$	$\frac{3}{2}$	$\frac{4}{2}$	$\frac{5}{2}$	$\frac{6}{2}$	$\frac{7}{2}$
4 5 6 7 8	$\frac{1}{3}$	$\frac{2}{3}$	$\frac{3}{3}$	$\frac{4}{3}$	$\frac{5}{3}$	$\frac{6}{3}$	$\frac{7}{3}$
5 6 7 8 9	$\frac{1}{4}$	$\frac{2}{4}$	$\frac{3}{4}$	$\frac{4}{4}$	$\frac{5}{4}$	$\frac{6}{4}$	$\frac{7}{4}$
6 7 8 9 10	$\frac{1}{5}$	$\frac{2}{5}$	$\frac{3}{5}$	$\frac{4}{5}$	$\frac{5}{5}$	$\frac{6}{6}$	$\frac{6}{7}.$

The list of numbers can be arranged in the following way, based on the rational listings: $\frac{1}{1}$ $\frac{1}{2}$ $\frac{2}{1}$ $\frac{3}{1}$ $\frac{1}{3}$ $\frac{1}{4}$ $\frac{2}{5}$ $\frac{3}{5}$ $\frac{4}{1}$ $\frac{5}{1}$.. with each rational number listed once only which shows that rational numbers are countable. This proves that the positive rational numbers are countable..

Note: George Cantor introduced his proof of the countability of sets on 1879. The method is used extensively in **Logic**, and in the **Theory of Computation**.

1.6   Exercise - 2

1. List the members of the set: $\{x/x \text{ is } +ve \text{ integer } < 8\}$.

2. Use builder notation to describe the set of : 0, 2, 4.

3. Determine if the first statement is the subset of the second:

a. The set of flying squirrels, the set of living creatures that can fly.

b. Set of Math teachers, set of all teachers.

In each of the following sets find their members explicitly:

4. $\{x/x \text{ is a real number, and } 2x - 3 = -4(x + 1) = 0\}$.

5. $\{n/n \text{ is an integer, and } 3(4n - 1) = 4n + 2\}$.

6. $\{x/x \text{ is a real number, and } 2x^2 - 2x - 3 = 1\}$.

7. $\{x/x \text{ is a real number, and } 4x^2 - 6x + 1 = 0\}$.

8. Suppose:

$$A \;=\; \{1, 3, 5\}.$$

$$B = \{1, 3\}.$$
$$C = \{3, 5\}.$$

Determine their subsets.

9. State if True, or False:

    a. $0 \in \phi$.

    b. $0 \subset \phi$.

    c. $\{\phi\} \subseteq \{\phi\}$.

    d. $\phi \in \{0\}$.

10. Use Venn-Diagram to show: $A \subseteq B, B \subseteq C$.

11. Translate and find the truth value:

a. $\forall y \in R\{y^2 \neq -1\}$.

b. $\exists x \in Z(x^2 = 2)$.

c. $\forall y \in Z(y^2 > 0)$.

d. $\forall x \in R(x^2 \in Z)$.

1.7   Infinite Numbers

A non-empty set S of real numbers is unbounded above if it has no upper bound, or unbounded below if it has no lower bound. Thus for points at infinity , we write:

$$supS = \infty \quad unbounded\ above \qquad (1.48)$$

$$infS = -\infty \quad unbounded\ below. \qquad (1.49)$$

Example:
if $S = \{x \mid x < 3\}$, then $supS = 3$, if $infS = -\infty$.
If $S = \{x \mid x \geq -3\}$, then $supS = \infty$, if $infS = -3$.
If S is the set of all integers, then: $supS = \infty$, and $infS = -\infty$.

The extended real numbers $\infty, -\infty$ possess the following arithmetic relationships which can be found in most text books:
a) If a is any real number, then:

$$a + \infty \;=\; \infty + a = \infty \qquad (1.50)$$

$$a - \infty \;=\; -\infty + a = -\infty \qquad (1.51)$$

$$\frac{a}{\infty} = \frac{a}{-\infty} = 0 \qquad (1.52)$$

b) If $a > 0,$, then:

$$a.\infty \;=\; \infty.a = \infty. \qquad (1.53)$$

$$a.(-\infty) \;=\; (-\infty).a = -\infty. \qquad (1.54)$$

c) If $a < 0$, then:

$$a.\infty \;=\; \infty.a = -\infty. \qquad (1.55)$$

$$a(-\infty) \;=\; (-\infty) = \infty. Also: \qquad (1.56)$$

$$\infty + \infty \;=\; \infty i \infty = (-\infty)(-(\infty)) = \infty. \qquad (1.57)$$

$$-\infty - \infty \;=\; \infty(\infty) = (-\infty)(\infty) = -\infty. And \qquad (1.58)$$

$$\mid \infty \mid \;=\; \mid -\infty \mid = \infty. \qquad (1.59)$$

1.8 Mathematical Induction

In the development of the real number system, the natural numbers become larger numbers, and emerge in the real number system as **positive integers** where $I = \{1, 2, ..\}$.

The positive integers have certain elementary properties which we list below, students are familiar with these properties:

1. The positive integers are positive numbers.
2. If n is a positive integer, then $n \geq 1$.
3.

$$
\begin{aligned}
2 &= 1 + 1, \\
3 &= 2 + 1, \\
4 &= 3 + 1, \\
5 &= 4 + 1, \\
&\quad . \\
&\quad . \\
n &= m + 1.
\end{aligned}
$$

where m = n-1.

4. $0 < 1 < 2 < 3 < 4...$
5. the sum of two-positive integers is positive integer. Also the product of two-positive integers is positive integer.
6. If m and n are positive integers, such that $m < n$, then $n - m$ is a positive integer too.
7. If n is a positive integer, then a positive integer m such that $n < m < n + 1$ does not exist.

An important property of the positive integers, is the mathematical induction.

Axiom of Induction:

If P is a set of positive integers with two properties:

1. P contains number 1.

2. Whenever P contains the positive integer n it also contains the positive integer $n + 1$, then n contains all the positive integers.

In proving statements that involve a positive integer n, it is frequently helpful to use the following principle.

Principle of Mathematical Induction:
Let S_n be a statement about the positive integer n. Suppose that :
1. S_1 is true.
2. S_{k+1} is true whenever S_k is true.
Then S_n is true for all positive integers n.
This is reasonable because, since S_1 is true, it follows from condition (2) (with $k = 1$) that S_2 is true. Then, using condition (2) with $k = 2$, we see that S_3 is true. Again using condition (2), this time with $k = 3$, we have that S_4 is true. This procedure can be followed indefinitely.

1.8.1   Examples on Mathematical Induction

The following examples illustrate the principles and use of mathematical induction.
Example-1:
Prove by induction that :

$$1 + 2 + ... + m = \frac{m(m+1)}{2} \qquad (1.60)$$

For all m.

Proof:
If (1.60) is true, then adding $(m + 1)$ to both sides yield:

$$(1 + 2 + ... + m) + (m + 1) \;=\; (m + 1) + \frac{m(m+1)}{2}$$

$$= \frac{2(m+1) + m(m+1)}{2}$$

$$= \frac{2m + 2 + m^2 + m}{2}$$

$$= \frac{m^2 + 3m + 2}{2}$$

$$= \frac{(m+1)(m+2)}{2}$$

$$= \frac{(m+1)(m+2)}{2}.$$

Replacing m with $(m+1)$ gives (1.60). Hence (1.60) is true for all m.

Example-2:

Prove by Induction that:

$$1^2 + 2^2 + 3^2 + \dots + m^2 = \frac{m(m+1)(2m+1)}{6} \qquad (1.61)$$

For all n.

Proof: Let S_m be the given formula:

1. Then for $m = 1$, S_1 is true because

$$1^2 = \frac{1(1+1)(2.1+1)}{6} = \frac{2.3}{6} = 1 = 1^2 \qquad (1.62)$$

2. Assume that S_m is true; that is,

$$1^2 + 2^2 + 3^2 + \dots + m^2 = \frac{m(m+1)(2m+1)}{6} \qquad (1.63)$$

Then S_{m+1} gives,

$$1^2 + 2^2 + \dots + (m+1)^2 = (1^2 + 2^2 + \dots + m^2) + (m+1)^2$$

$$= \frac{m(m+1)(2m+1)}{6} + (m+1)^2$$

$$= (m+1)\frac{m(2m+1)+6(m+1)}{6}$$

$$= (m+1)\frac{2m^2+7m+6}{6}$$

$$= \frac{(m+1)(m+2)(2m+3)}{6}$$

$$= \frac{(m+1)[(m+1)+1][2(m+1)+1]}{6}.$$

Replacing k with $(m+1)$ gives,

$$= \frac{m[m+1][2m+1]}{6} \qquad (1.64)$$

So S_{m+1} is true. Then by the principle of Mathematical induction S_m is true for all m.

Example-3:
Let $a_1 = 1$, and $a_{m+1} = \frac{1}{m+1}a_m$, $m \leq 1$. Then find an explicit formula for a_m.

Solution:
Consider $m = 1, 2, 3, ..$ then:

$$a_1 = 1 = \frac{1}{1}$$

$$a_2 = \frac{1}{2}a_1 = \frac{1}{1.2}.1 = \frac{1}{1.2}$$

$$a_3 = \frac{1}{3}a_2 = \frac{1}{3}.\frac{1}{1.2} = \frac{1}{1.2.3}$$

.

.

.

$$a_n = \frac{1}{1.2.3...m} = \frac{1}{m!}, \; then, \; \rightarrow a_{m+1} = \frac{1}{(m+1)!}.$$

Example-4:

Let $p_m = 3m + 16 > 0$. If p_m is true , so is p_{m+1}.

Proof:

$$
\begin{align}
p_{n+1} &= 3(m+1) + 16 && (1.65) \\
&= 3m + 3 + 16 && (1.66) \\
&= (3m + 16) + 3 > 0 + 3. && (1.67)
\end{align}
$$

For smallest $m_0 \implies 3m_0 + 16 = 0 \implies 3m_0 = -16, \implies$ $m_0 = -5$.

therefore, p_m is true for $m \geq -5$.

1.8.2 Important Definition on Point Sets

1. Neighborhoods:

A neighborhood of point a is the set of all points x such that $\mid x - a \mid < \epsilon$, where $\epsilon > 0$, and from the property of absolute values x consists of all points satisfying,

$a - \epsilon < x < a + \epsilon$, or

$-\epsilon < x - a < \epsilon$.

It consists of all points whose distance from a is less that ϵ. The point a is the midpoint of its neighborhood. It is possible to avoid the concept of neighborhood by replacing it by an open interval or open sphere in higher spaces.

2. Interior Points:

A point $a \in S$ is called an interior point of S if there exist a δ neighborhood of a all of whose points belong to S. Alternatively, a is an interior point of S if there exist an open interior $I \subset S$ such that $a \in I$.

the set of interior points $S$is called the interior of S.

3. A set is said to be open if each of its points is an interior point.

4. A set is called closed set if it contains all its limit points.

1.9 Exercise - 3

Prove the assumptions in 1, and 2 by induction:

$$\text{1.} \quad 1^2 + 2^2 + . + . + x^2 = \frac{x(x+1)(2x+1)}{6} \qquad (1.68)$$

$$\text{2.} \quad 1^2 + 3^2 + . + . + (2x+1)^2 = \frac{x(4x^2-1)}{3}. \qquad (1.69)$$

3. Find and prove by induction an explicit formula for a_n if $a_1 = 1$ and for $n \geq 1$.

$$a) \quad a_{n+1} = \frac{a_n}{(n+1)(2n+1)} \qquad (1.70)$$

$$b) \quad a_{n+1} = \frac{3a_n}{(2n+2)(2n+3)} \qquad (1.71)$$

$$c) \quad a_{n+1} = \frac{2n+1}{n+1}a_n \qquad (1.72)$$

$$d) \quad a_{n+1} = (1+\frac{1}{n})^n a_n. \qquad (1.73)$$

4. Let p_m be the proposition that:

$$1 + 2 + ... + m = \frac{(m+2)(m-1)}{2}. \qquad (1.74)$$

a) Show that p_m implies p_{m+1}.

b) Is there an integer m for which p_m is true?.

5. For what values of m is:

$$\frac{1}{m!} > \frac{8^m}{(2m)!}?$$

(1.75)

Prove your answer by induction.

6. Prove:

$$\sum_{i=1}^{m} i^2 = [\frac{m(m+1)}{2}]^2.$$

(1.76)

7. Prove,

$$\sum_{i=1}^{m} c = mc.$$

(1.77)

Chapter 2

Functions and Limits

Definition-1: Intuitively we can define the function as a plane graph, but since a plane graph is formed out of set of points, and each point is given by a pair of numbers, then this will lead to the following, more proper, definitions.

Definition-2: A function f from (on) a set X to (onto) a set Y means a **Rule** which arrange to each $x \in X$ a unique element $f(x) \in Y$.

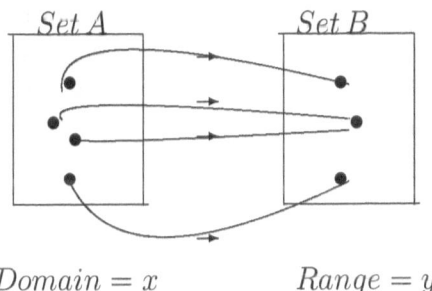

$Domain = x$ $Range = y$

As a collection pairs it can be written as $\langle x, f(x) \rangle$, where

$\langle x, f(x) \rangle \in (X, Y)$ is called the **Graph**
A subset $G \subset (X, Y)$ is the **graph of a function** on x IFF
$x \in X$ is the unique pair in G.

Definition-3: Let X, and B be any two sets. A function
from X into Y is a subset of $X \times Y$, and hence is a set of
ordered pairs $\langle x, y \rangle$ with the property that each $x \in X$ belongs
to precisely one pair $\langle a, b \rangle$.

Definition-4: Function is also defined as " Mapping",
Mapping f from a set X to a set Y is written as : $f : X \rightarrow Y$,
and is a rule which assigns to each $x \in X$ a unique element
$y \in Y$. x, y are the real variables, Instead of $\langle x, y \rangle \in f$ we
usually write $y = f(x)$. Then y is called the image of x under
f. The set X is called the domain of f. The range of f is
the set $\{y \in Y | y = f(x) for\ x\}$. That is the range of f is the
subset of Y consisting of all images of elements of X. Such a
function is sometimes called a mapping of X into y.
If $Z \subset Y$, then $f^{-1}(Z)$ (which is called inverse image of Z)
is defined as $\{x \in X | f(x) \in Z\}$, the set of all points in the
domain of f whose image are in Z.

Definition-5: Let D and R be two sets of objects with $x \in D$,
and $y \in R$. Then $y = f(x)$ is called a function with domain D
and range R if and only if to each member x of D there corresponds at least one member y of R, and for each member y of
R there is at least one member x of D to which y corresponds.
The general member of D is called the independent variable,
and the general member of R is called the dependent variable.
In case no two members of Rcorresponds to the same member
of D, $f(x)$ is called single-valued. In case R consists of just one
object, $f(x)$ is called a constant function. In case D consists
of real numbers, $f(x)$ is called a function of real variable. In
case R consists of real numbers , $f(x)$ is called real-valued.

The symbol $f(x_0)$ denotes the members of R that corresponds to the member x_0 of D. Or function can also be defined as input, output system.

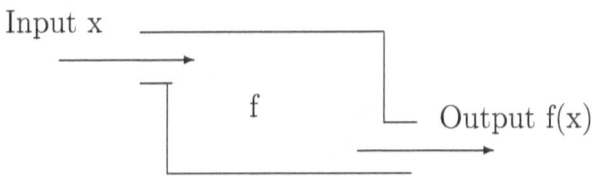

. . . .

Theorem:

If the function $f : X \to Y$ has an inverse function f^{-1} then,
1. the function f is a $1-1$ function on $X \to Y$.
2. There exist a function $g : Y \to X$, such that,

$$g\left(f(x)\right) = x, \ for \ all \ x \in X, and \qquad (2.1)$$
$$f\left(g(x)\right) = x, for \ all \ x \in Y. \qquad (2.2)$$

Example-1: The set $f = \{\langle x, x^3\rangle| - \infty < x < \infty\}$ is the function described by the equation,

$$f(x) = x^3 \ (-\infty < x < \infty). \qquad (2.3)$$

The domain and the range of f is the whole real line. In addition,

$$f(2) = 8, \qquad (2.4)$$
$$f^{-1}(8) = \{2\}, \qquad (2.5)$$
$$f\left(\{x|x^3 = 8\}\right) = \{8\}, \qquad (2.6)$$
$$f\left([0, 2)\right) = [0, 8). \qquad (2.7)$$

Example-2: The set $f = \{\langle x, x^2 \rangle | 0 \leq x \leq 1\}$ is the function described by the equation,
$f(x) = x^2, [0 \leq x \leq 1]$ is not a function since the correspondence function,
$g : x^2 \rightarrow x$ is not single-valued and hence undefined.

Example-3: Suppose $F = \{(x, y) : x \in R, and\, y = x^3\}$.
Then we can state the following:
a) F is a function for the following reasons: If $(x, y_1) \in F$, and $(x, y_2) \in F$
$\Rightarrow y_1 = x^3$, and $y_2 = x^3 \Rightarrow y_1 = y_2$.
b) $D_F = R$, for every member of D_F, and for every $x \in R$, x^3 is defined, so $x \in D_F$ and $F(x) = x^3$.
c) $R_F = \{y/y \geq 0\}$. For some $x \in R$, $y = x^3$. So $y \geq 0$.
Conversely, if $y \geq 0 \Rightarrow (\sqrt{y}, y) \in F$. So $y \in R_F$.

Note: The symbol $[x]$, where x is a real number, denotes a unique integer n such that: $n \leq x \leq n + 1$. The number $[x]$ is called the greatest integer, and $f(x) = [x]$, f is called the greatest integer function.

2.1   One-to-one Functions

Definition: A function F is one-to-one function if and only if (IFF) x_1,and x_2 are in domain of the function (D_F), and $F(x_1) = F(x_2)$, it follows that : $x_1 = x_2$.

Examples:

1. Let $F : F(x) = 5x - 3$. Then F is one-to-one, because:
$x_1 \Rightarrow F(x_1) = 5x_1 - 3$
$x_2 \Rightarrow F(x_2) = 5x_2 - 3$.. If $F(x_1) = F(x_2)$.
Then, $5x_1 - 3 = 5x_2 - 3$. This means $x_1 = x_2$.

2. Let,

$$F(x) = \frac{2x}{x+3}.$$

Then for x_1

$$F(x_1) = \frac{2x_1}{x_1+3}.$$

And for x_2

$$F(x_2) = \frac{2x_2}{x_2+3}.$$

If $F(x_1) = F(x_2)$. Then,

$$\frac{2x_1}{x_1+3} = \frac{2x_2}{x_2+3}.$$

Simplifying, gives:

$$2x_1(x_2+3) = 2x_2(x_1+3)$$
$$2x_1x_2 + 6x_1 = 2x_1x_2 + 6x_2$$
$$6x_1 = 6x_2$$
$$\Rightarrow x_1 = x_2.$$

3. Let $h(r) = r^2 + 2r$ then,
$r_1 \Rightarrow h(r_1) = r_1{}^2 + 2r_1$
$r_2 \Rightarrow h(r_2) = r_2{}^2 + 2r_2$.
If $h(r_1) = h(r_2)$. Then, $r_1^2 + 2r_1 = r_2^2 + 2r_2$
$\Rightarrow (r_1 - r_2)(r_1 + r_2) + 2(r_1 - r_2) = 0$
From this we get: $r_1 - r_2 = 0$, $\Rightarrow r_1 = r_2$.
Or $r_1 + r_2 + 2 = 2r_1 = 2 = 0 \Rightarrow r_1 = 1$.
If $r_1 = 1$, then $r_1 + r_2 + 2 = 0$ gives $r_2 = -3$.
Then $F(r_1) = F(1) = 3$, and $F(r_2) = F(-3) = 3$
Thus, $F(r_1) = F(r_2)$.

2.1.1   Properties of Functions

Here, we are going to list the properties only without a proof, because it is similar to the proof in real numbers.

Definition: Let f and g be real functions, and
$\{x : x \in D_f$, and $x \in D_y$ $\}$, then:
i) $f + g = \{(x, y) : y = f(x) + g(x)\}$Addition Rule.
ii) $f - g = \{(x, y) : y = f(x) - g(x)\}$ Subtraction rule.
iii) $f.g = \{(x, y) : y = f(x)g(x)\}$ Multiplication rule.
iv) $f/g = \{(x, y) : y = f(x)/g(x)\}$ Quotient rule .

2.2   Invertible Functions

Definition: If f is one-to-one function , then the function $\{(x, y)/(x, y) \in f\}$ is called the inverse of f, and is denoted as f^{-1}.

Theorem:
i) A function f is invertible if and only if , for every element $y \in R_f$. there is a unique element $x \in D_f$, such that $f(x) = y$.
ii) If f is invertible, then $D_f^{-1} = R_f$, and for each $y \in D_f^{-1}$, $f^{-1}(y)$ is the unique solution x of the equation $f(x) = y$.

Examples:

1. Show if f is invertible , $f(x) = 3x - 4$.
$\Rightarrow y = f(x) = 3x - 4$
Solving for x, we get $x = \frac{1}{3}(y + 4)$ valid for any y.
Thus by (i) f is invertible. Then $f^{-1}(y) = \frac{1}{3}(y + 4)$.
Therefore, $f^{-1}(y) - \frac{1}{3}(y + 4)$.

2. Show if the function $f : f(t) = \frac{t}{t+1}$ is invertible.

Let $s = f(t)$, then $s = \frac{t}{t+1}$.

Solving for t gives:

$t = st + t$, then $t = \frac{s}{1-s}$, where $s \neq 1$.

If $s = 1$, then $1 = \frac{t}{t+1}$, and t has no solution.

Thus, $R_f = \{s : s \in R, and s \neq 1\}$.

Therefore, for each $s \in R_f$, there is a unique solution $t = \frac{s}{1-s}$ of equation $s = f(t)$. Therefore, f is invertible. Then f^{-1}:

$f^{-1}(s) = \frac{s}{1-s}$.

2.3   Limits of Functions

A number l is said to be the limit of $f(x)$ as x approaches a, if for every $\epsilon > 0$, there exist a $\delta > 0$ such that, $\mid f(x) - l \mid < \epsilon$, whenever $0 < \mid x - a \mid < \delta$.

In such case we write: $lim_{x \to a} f(x) = l$.

Theorems:

1. If a limit exists it is unique.

2. If $lim_{x \to a} f_1(x) = l_1$, $lim_{x \to a} f_2(x) = l_2$, then,

$$
\begin{array}{rcll}
(a) \ lim_{x \to a}[f_1(x) + f_2(x)] & = & l_1 + l_2 & (2.8) \\
(b) \ lim_{x \to a} f_1(x).f_2(x) & = & l_1.l_2 & (2.9) \\
(c) \ lim_{x \to a}[f_1(x)/f_2(x)] & = & l_1/l_2, if \ l_2 \neq 0. & (2.10)
\end{array}
$$

In the above definitions of limits, if $0 < \mid x - a \mid < \delta$ is replaced with $0 < x - a < \delta$ then l is called the right hand limit and is written as $lim_{x \to a+} f(x) = l$, and left hand limit is written as $lim_{x \to a-} f(x) = l$. If the right and left hand limits are equal, then the limit exist.

definition: If $s_n = c$ for $n \geq k$, then $\mid s_n - c \mid = 0$ for $n \geq k$, and $\lim_{n \to \infty} s_n = c$.

A Simple Approach to the Idea of Limits:

To explain the limits of functions stated above in simpler form, we will consider a simple linear function such as: $f(x) = 3x + 2$, with the graph shown below.

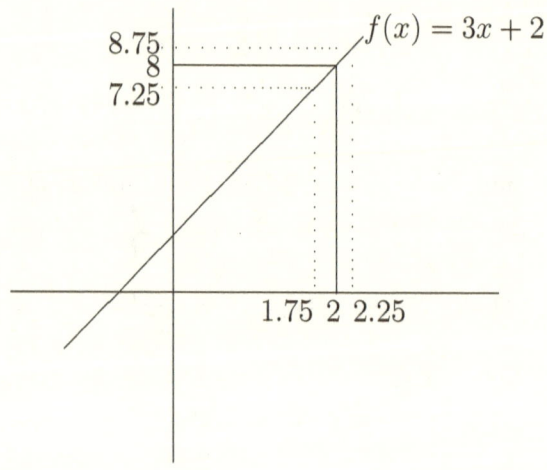

We need to prove that :

$$\lim_{x \to 2} f(x) = 8.$$

We will consider a boundary of ∓ 0.75 for the function $f(x)$, that is : $8 \pm .75$ or,

$$7.25 < f(x) < 8.75$$
$$7.25 < 3x + 2 < 8.75$$
$$5.25 < 3x < 6.75$$
$$1.75 < x < 2.25.$$

To see how close we can get to $x = 2$, then,

$$-.25 < x - 2 < .25.$$

The last expression is just the absolute value,

$$|x - 2| < 0.25.$$

With the corresponding value for $f(x)$:

$$|f(x) - 8| < 0.75.$$

This proves that the limit of $f(x)$ is equal to 8 as $x \to 2$. Suppose we chose a smaller boundary for the function $f(x)$ such as ± 0.5, i.e. $8 \pm .5 = (7.5 - 8.5)$. As shown in the graph:

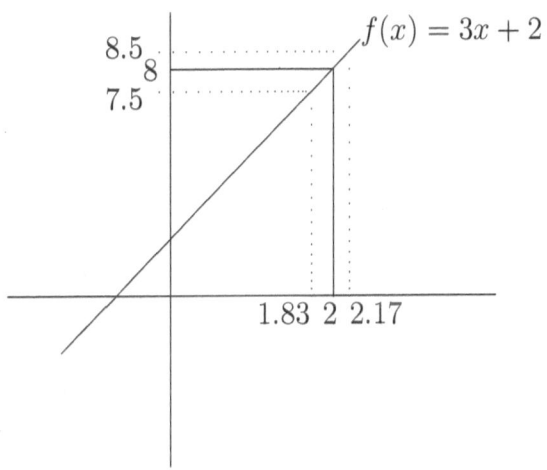

Following the same steps, as before:

$$7.5 < f(x) < 8.5$$
$$7.5 < 3x + 2 < 8.5$$
$$5.5 < 3x < 6.5$$
$$1.83 < x < 2.17.$$

To see how close we can get to $x = 2$, then,

$$-.17 < x - 2 < .17.$$

The last expression is just the absolute value,

$$|x - 2| < 0.17.$$

With the corresponding value for $f(x)$:

$$|f(x) - 8| < 0.5.$$

We can continue with the same pattern taking smaller and smaller boundaries. Suppose we use a general pattern by taking the boundary of $\pm\epsilon$, then following the same steps we get:

$$8 - \epsilon < f(x) < 8 + \epsilon$$
$$8 - \epsilon < 3x + 2 < 8 + \epsilon$$
$$6 - \epsilon < 3x < 6 + \epsilon$$
$$2 - \frac{\epsilon}{3} < x < 2 + \frac{\epsilon}{3}.$$

To see how close we can get to $x = 2$, then,

$$-\frac{\epsilon}{3} < x - 2 < \frac{\epsilon}{3}.$$

The last expression is just the absolute value,

$$|x - 2| < \frac{\epsilon}{3}.$$

With the corresponding value for $f(x)$:

$$|f(x) - 8| < \epsilon.$$

If we chose $\frac{\epsilon}{3} = \delta$. Then we can say:

$$|f(x) - 8| < \epsilon, \ when \ |x - 2| < \delta.$$

Examples on Limits of Supremum and Infimum

For the given sequence: $2, -2, 1, -1, 1, -1, 1, -1, ..$

Find the following:

a) l.u.b

b) g.l.b

c) lim sup or $(\overline{\lim})$

d) lim inf or $(\underline{\lim})$

Solution:

$$\downarrow \quad -2 \quad \downarrow \quad -1 \quad \quad 1 \quad \downarrow \quad 2 \quad \downarrow$$
$$\text{-2-}\epsilon \qquad \text{-2+}\epsilon$$

Or using line of numbers to show the sequence of numbers:

a) The l.u.b $= 2$, since all the terms are ≤ 2, while at least one term (the first) is greater than $2 - \epsilon$ for any $\epsilon > 0$.

b) The g.l.b $= $ -2, since all the terms are ≥ -2 while at least one term (the second) is less than -2ϵ for any $\epsilon > 0$.

c) The lim Sup or $(\overline{\lim}) = 1$. Since infinitely many terms of the sequence are greater than (normally all 1's) $1 - \epsilon$, for $\epsilon > 0$, while only a finite number of terms are greater than $1 + \epsilon$ (normally first term).

d) lim Inf or $(\underline{\lim}) = $ -1. Since infinitely many terms are less than $-1 + \epsilon$ for any $\epsilon > 0$ (namely all -1's), while only a finite number of terms are less than $-1 - \epsilon$ for any $\epsilon > 0$ (namely the second term).

2.3.1 Uniqueness of the Limit

Theorem: The limit of a convergent sequence is unique.
proof: Suppose that,
$\lim_{n \to \infty} S_n = K$, , and $\lim_{n \to \infty} S_n = L$
We must show that $K = L$ to prove the uniqueness.
Suppose for $\epsilon > 0$, there are integers N_1 and N_2 such that,

$$\mid S_n - K \mid < \epsilon \quad \text{if } n \geq N_1 \text{ , and } \quad \mid S_n - L \mid < \epsilon \quad \text{if } n \geq N_2,$$

these inequalities both hold if $n \geq N = max(N_1, N_2)$. Which implied that:

$$\mid K - L \mid = \mid (K - S_N) + (S_N - L) \mid < \epsilon + \epsilon = 2\epsilon.$$

We conclude that $\mid K - L \mid = 0$ for every $\epsilon > 0$, thus $K = L$.

Infinite Limits:

If $f(x)$ is a function defined in an open interval $I = (a, b)$, taking the limit we get:

$$a) \; limit_{x \to a} f(x) = \infty.$$

Which means for some positive number k, there exist a positive number δ such that:

$$If \; 0 < |x - a| < \delta \Rightarrow, \; then \; f(x) > k,$$

and,

$$b) \; \lim_{x \to a} f(x) = -\infty.$$

Which means for some positive number δ :

$$If \; 0 < |x - a| < \delta \Rightarrow, \; then \; f(x) < -k.$$

Example: Find

$$\lim_{x\to1}\frac{3}{(x-1)^2}.$$

Solution: As $x \to 1$ the limit is equal ∞. But to show that (a) is satisfied, we choose a positive number k such that:

$$k > 0 \Rightarrow \frac{3}{(x-1)^2} > k.$$

$$3 > (x-1)^2 K.$$

$$(x-1)^2 < \frac{3}{k}.for\ x \neq 1.$$

And this is satisfied if:

$$|x-1| < \sqrt{\frac{3}{k}}.$$

If we let $\delta = \sqrt{\frac{3}{k}}$, we get:

$$0 < |x-1| < \delta.$$

$$then,\ \frac{3}{(x-1)^2} > k.$$

Which proves that:

$$\lim_{x\to}\frac{3}{(x-1)^2} = \infty.$$

2.4 Exercise - 1

1. Describe the following functions:

a) $\{(x,y)/x \in R, and\ y - 3x + 1 = 0\}$
b) $\{(u,v)/u \in R, v \in R,\ and\ u^2 + v^2 = 4\}$

c) $\{(x, y)/x \le 1, y = x + 2, \; or \; x > 1, \; and \; y = 3\}$.

2. Show the following functions are one-to-one functions:

a) $F(x) = 3x - 2$.
b) $g(x) = -\frac{1}{2}x + 3$.
c) $h(x) = x^3, x \le 0$.
d) $J(x) = \frac{1}{x}, x > 0$.

3. Find the inverse of each of the following functions:
a) $f(x) = \frac{3x}{2x+1}$.

b) $g(x) = \frac{x}{1-x^2}$, $x > 0$.

2.5   Derivative of Functions

The derivative of $f(x)$ is defined as ,

$$Df(x) = f'(x) = \lim_{h \to 0} \frac{f(x+h) - f(x)}{h}, \qquad (2.11)$$

and this may or may not exist depending on the existence of the limit. Here we will define four quantities for the derivative,

$$D^+ f(x) \;\; = \;\; \lim_{h \to 0^+} \frac{f(x+h) - f(x)}{h} \qquad (2.12)$$

$$D^- f(x) \;\; = \;\; \lim_{h \to 0^-} \frac{f(x+h) - f(x)}{h} \qquad (2.13)$$

$$D_+ f(x) \;\; = \;\; \lim_{h \to 0^+} \frac{f(x+h) - f(x)}{h} \qquad (2.14)$$

$$D_- f(x) \;\; = \;\; \lim_{h \to 0^-} \frac{f(x+h) - f(x)}{h}. \qquad (2.15)$$

Which are either finite, positively infinite, or negatively infinite.

In case, $D^+f(x) = D_+f(x)$, then $f(x)$ has a right hand derivative denoted by $f'_+(x)$.

In case, $D^-f(x) = D_-f(x)$, then $f(x)$ has a left hand derivative denoted by $f'_-(x)$.

In case , $D^+f(x) = D_+f(x) = D^-f(x) = D_-f(x)$, then $f(x)$ has a derivative $f'(x)$. Conversely, if $f(x)$ has a derivative, then all the above 4-derivatives are equal.

Definition:

Let f be a real-valued function on an interval $J \subset R'$ if $c \in J$ we say that f has a derivative at c,

$$\lim_{x \to c} \frac{f(x) - f(c)}{x - c}, \qquad (2.16)$$

if this limit exist it is denoted by $f'(x)$.

If $f'(x)$ exist, then f is continuous at c (the converse is false). If f is not continuous at any point in J then f can not have a derivative at any point in J. It is also possible that $f'(c)$ does not exist even though f is continuous at c.

Theorem: If f is differentiable at a, then f is continuous at a.

Proof: The proof is to show that $\lim_{x \to a} f(x) = f(a)$, and this is done by showing that $f(x) - f(a) = 0$:

From the given information we know that:

$$f'(a) = \lim_{x \to a} \frac{f(x) - f(a)}{x - a} \quad exist.$$

Since ,

$$f(x) - f(a) = \frac{f(x) - f(a)}{(x - a)}(x - a). \qquad (2.17)$$

Thus, using the limit and product rule we can write,

$$\lim_{x \to a}[f(x) - f(a)] = \lim_{x \to a} \frac{f(x) - f(a)}{(x - a)}(x - a) \qquad (2.18)$$

$$= \lim_{x \to a} \frac{f(x) - f(a)}{(x - a)} \cdot \lim_{x \to a} (x - a) \quad (2.19)$$

$$= f'(a).0 = 0. \quad (2.20)$$

Also,

$$\lim_{x \to a} f(x) = \lim_{x \to a} [f(a) + (f(x) - f(a))] \quad (2.21)$$

$$= \lim_{x \to a} f(a) + [f(x) - f(a)] \quad (2.22)$$

$$= f(a) + \lim_{x \to a} [f(x) - f(a)]. \quad (2.23)$$

$$= f(a) + 0. \quad (2.24)$$

Therefore, f is continuous at a.

Note: The converse of this theorem is false; there are functions that are continuous but not differentiable:

For example: the function $f(x) = |x| = 0 = f(0)$ is continuous at 0 because, $lim_{x \to 0} f(x) = \lim_{x \to 0} |x| = 0 = f(0)$. But f is not differentiable at 0. In fact for the function $f(x) = |x|$ the formula for the derivative $f'(x)$ is given by:

$$f'(x) = \begin{cases} 1 \ if x > 0 \\ -1 \ if x < 0, \end{cases}$$

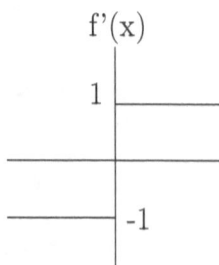

Note: The fact that $f'(0)$ does not exist is reflected geometrically in the fact that the curve $y = |x|$ does not have the tangent line at $(0,0)$.

Important theorems:
Theorem-1 : If a function is monotonic, then it has a derivative almost everywhere.

Theorem-2 : If a function is of bounded variation, then it has derivative almost everywhere.

Theorem-3 : If $\sum_{k=1}^{\infty} f_k(x)$ is a series of functions of bounded variation which converges to $S(x)$ in $[a, b]$, then almost everywhere in $[a, b]$,

$$S'(x) = \sum_{k=1}^{\infty} f'_k(x) \tag{2.25}$$

2.5.1 Another Definition for Derivative

From the previous standard definition of the derivative ,

$$f'(a) = \lim_{x \to a} \frac{f(x) - f(a)}{x - a}, \tag{2.26}$$

it is understood to mean that, $\frac{[f(x) - f(a)]}{[x-a]}$ is a good approximation to $f'(x)$ that gets better as x gets closer and closer to a. Now we will introduce the actual difference between derivative and the average rate of change,

$$f'(a) = \frac{f(x) - f(a)}{x - a} + E(x, a), \tag{2.27}$$

where $E(x, a)$ is the error. Here we say $f'(x)$ is the value of the derivative provided the error $E(x, a)$ gets closer to 0 as x gets closer to a. We can make the absolute value of our error as small as it needs to be by controlling the distance between x and a.

We fix a real number a and say f is differentiable at $x = a$ if and only if there is a number, denoted by $f'(a)$, such that the error,

$$E(x, a) = f'(a) - \frac{f(x) - f(a)}{x - a}, \qquad (2.28)$$

satisfies the following condition:

For any satisfied bound on this error, $\epsilon > 0$, we can find a distance, δ, such that if x is within δ of a $(0 < | x - a | < \delta)$, then the error size within the allowable bound is $(|E(x, a)| < \epsilon)$. We call $f'(a)$ the derivative of f at a. The key to this definition is to emphasis on the error or difference between this derivative and the average rate of change. We have differentiability when this error can be made as small as desired by shrinking the distance between x and a.

A note about ϵ and δ:
The constraint is that ϵ which represents the size of the allowable error $E(x, a)$, must be positive($\epsilon > 0$), and since δ represent a distance between x, and a point say a then $\delta > 0$.

Example: For the given function $f(x) = 3x^2 - 5x + 2$, suppose we want to check the differentiability of $f(x)$ at some point say 2.
Solution: We differentiate $f(x)$, then substitute $x = 2$:
$f'(x) = 6x - 5$, and at $x = 2 \rightarrow f'(x) = 6(2) - 5 = 7$.
Then the error term E at 2 is, using equation (2.36).

$$
\begin{aligned}
E(x, 2) &= 7 - \frac{f(x) - f(2)}{x - 2}. \\
&= 7 - \frac{3x^2 - 5x + 2 - 4}{x - 2}.
\end{aligned}
$$

$$= 7 - \frac{3x^2 - 5x - 2}{x - 2}.$$

$$= \frac{(3x + 1)(x - 2)}{(x - 2)}.$$

$$= 7 - (3x + 1).$$

$$Or \ E(x, 2) = -3x + 6.$$

where the error at any point a is ,

$$E(x, a) = f'(a) - \frac{f(x) - f(a)}{x - a} \qquad (2.29)$$

If we choose ϵ to be a very small number such as: $\epsilon = 0.01$ this means we are looking at:

$$| \ error \ | = | -3x + 6 \ | < 0.01, \qquad (2.30)$$

Or,

$$-.01 < -3x + 6 < .01.$$
$$-6.01 < -3x < -5.99.$$
$$then, \ 1.996 < x < 2.003.$$
$$And, \ 1.997 - 2 < E(x, 2) < 2.003 - 2.$$
$$-0.003 < E(x, 2) < 0.003.$$

Then, $|E(x, 2)| < .003 < \epsilon.$

What Does $f\prime(x)$ say about $f(x)$?

Consider the curve $y = f(x)$, since $f'(x)$ represents the slope of the curve at the point $(x, f(x))$, then it tells us the direction of the curve at each point.It will tell us if the function of the curve is increasing i.e. $f'(x) > 0$, or the function of the curve is decreasing i.e. $f'(x) < 0$. So it appears that f increases when $f'(x)$ is positive, and decreases when $f'(x)$ is negative.

The general important results can be stated as follows:
If $f(x) > 0$ on an interval, then f is increasing on that interval.
If $f(x) < 0$ on an interval, then f is decreasing on that interval.

What Does $f\prime\prime(x)$ say about f(x) ?

Here we will state the following important results:
If $f\prime\prime(x) > 0$ on an interval, then f is concave upward on that interval.
If $f\prime\prime(x) < 0$ on an interval, then f is concave downward on that interval.

2.6   Exercise - 2

Find the derivative of the functions using the definition of derivative. State the domain of the function and the domain of its derivative.

1. $f(x) = \sqrt{1 + 2x}$

2. $f(x) = \frac{4x}{x+1}$

3. The unemployment rate $U(t)$ varies with time. The table below gives the percentage of unemployed in the U.S. labor 1993 to 2002.

t	$U(t)$	t	$U(t)$
1993	6.9	1998	4.5
1994	6.1	1999	4.2
1995	5.6	2000	4.0
1996	5.4	2001	4.7
1997	4.9	2002	5.8

a) What is the meaning of $U'(t)$? What are its units?
b) Construct a table for the values of $U'(t)$.

4. The table gives population densities for a type of birds (per acre):
a) Describe how the rate of change of population varies.
b) Estimate the inflection points of the graph. What is the significance of these points?

t	$P(t)$	t	$P(t)$
1927	0.1	1930	0.6
1932	2.5	1934	4.6
1936	4.8	1938	3.5

Chapter 3

Sequence and Limits

3.1 Sequence

A function whose domain of definition is an ordered set similar to the set of positive integers (I^+) in their natural order is called **sequence**, thus a sequence is a function whose value can be written as $f(n)$ or f_n . But usually a sequence $x_1, x_2, x_3, ...$ can be represented by $\{x_n\}$, and its subsequence $x_{n_1}, x_{n_2}, x_{n_3}, ...$ is represented by $\{x_{n_i}\}$. Thus $\{x_{n_i}\}$ is called the subsequence of the original sequence $\{x_n\}$.

definition: A sequence is a function whose domain is the set of all real numbers.

To get a better idea about the sequence we will look at the simple primitive example; Suppose we want to express the familiar fraction $\frac{22}{7}$ as a decimal, first we will perform the long division steps as follows: divide 22 by 7, the result is : 3, and remainder 1, then continue the long division on step 2, starting with the remainder 1.0 divided by 7 the result is 3.1 with the remainder of 3, and so on, then list the sequence of partial results as follows:

$$3 \rightarrow 3.1 \rightarrow 3.14 \rightarrow 3.142 \rightarrow 3.1428 \rightarrow 3.14285 \rightarrow 3.142867$$

From this we can get infinite sequence of decimal approximations as shown:

$$
\begin{aligned}
a_1 &= 3 \\
a_2 &= 3.1 \\
a_3 &= 3.14 \\
a_4 &= 3.142 \\
a_5 &= 3.1428 \\
a_6 &= 3.14285 \\
a_7 &= 3.142857
\end{aligned}
$$

.

.

Even though the above sequence looks like it does not end, but it does have a definite end. Then we could say that the fraction $\frac{22}{7}$ is the limit of the sequence.

$$a_n = f(n) = \{f(n)\}_{n=1}^{\infty} = \lim a_n \qquad (3.1)$$

Definition of Sequence:
A sequence $S = \{s_n\}_{n=1}^{\infty}$ of real numbers is a function form $(I^+)$the set of positive integers, into R, the set of real numbers. Or symbolically we write: $S = f : I^+ \rightarrow R$. Intuitively we use sequence to mean a function with domain (I^+)

Example:

Here we will use the familiar example from Calculus:
Consider the sequence :

$$\{a_n\}_{n=1}^{\infty} = \{(1 + \frac{1}{n})^n\}_{n=1}^{\infty}. \tag{3.2}$$

If we take the terms we get:

$$a_1 = (1 + \frac{1}{1})^1 = 2$$

$$a_2 = (1 + \frac{1}{2})^2 = 2.25$$

$$a_3 = (1 + \frac{1}{3})^3 = 2.37037037037$$

$$a_4 = (1 + \frac{1}{4})^4 = 2.44140625000$$

$$a_5 = (1 + \frac{1}{5})^5 = 2.4883200000$$

.

.

.

$$a_{10} = (1 + \frac{1}{10})^{10} = 2.59374246010$$

.

.

.

$$a_{1000000} = (1 + \frac{1}{10^6})^{10^6} = 2.718280469$$

In fact it is known that:

$$lim_{n \to \infty}(1 + \frac{1}{n})^n \simeq e \simeq 2.718281828, \tag{3.3}$$

is a sequence that converges to e.

Definition of Subsequence:

If $S = \{s_n\}_{n=1}^{\infty}$ is a sequence of real numbers and $B = \{s_i\}_{i=1}^{\infty}$ is a subsequence of sequence of positive integers (I^+), then the composite function $S \circ B$ is called a subsequence of S.
Note that for $i \in I^+$ we have :

$$B(i) \;=\; n_i, \tag{3.4}$$
$$S \circ B(i) \;=\; S[B(i)] = S(n_i) = s_{n_i}, hence \tag{3.5}$$
$$S \circ B \;=\; \{s_{n_i}\}_{i=1}^{\infty}. \tag{3.6}$$

Example:
Let $S = 1, 0, 1, 0, \dots$ be the sequence, with $B = \{n_i\}_{i=1}^{\infty}$ defined by,
$n_i = 2i - 1, \;\; (i \in I)$.
So the : $n_1 = 1, n_2 = 3, n_3 = 5, \dots.$ Then

$$S \circ B = S(B) = 1, 1, 1, \dots. = the\ subsequence\ of\ S. \tag{3.7}$$

Example:
Let $C = \{c_n\}_{n=1}^{\infty} = \{\sqrt{n}\}_{n=1}^{\infty}$ and $N = \{n_i\}_{i=1}^{\infty} = \{i^4\}_{i=1}^{\infty}$

Then $C \circ N = C(N) = \{c_{n_i}\}_{i=1}^{\infty} = \left\{\sqrt{i^4}\right\}_{i=1}^{\infty} = \{i^2\}_{i=1}^{\infty}.$

Which is the subsequence of C.

Note: A set of numbers may form a sequence if arranged in one way, but not if arranged in some other way. Thus the set of natural numbers $N = \{\dots, -2, -1, 0, 1, 2, \dots\}$ form sequence if arranged in the order of rank. However, the numbers: $\underbrace{2, 4, 6, 8,}\ \underbrace{9, 11, 13, 17, \dots}$ is not a sequence since the first succession is of even integers followed by the succession of odd integers, because this arrangement does not have a one-to-one correspondence with the positive integers which preserve the order of arrangement of numbers.

3.2 Infinite sequence:

An infinite sequence of real numbers is a real-valued function defined on a set of integers $\{n \mid n \geq k\}$. The terms of sequence can be listed as:

$$\{S_n\}_k^\infty = \{s_k, s_{k+1}, ...\}. \tag{3.8}$$

For example,

$$\{(-1)^n\}_0^\infty = \{1, -1, 1, ..., (-1)^n, ..\}, \tag{3.9}$$

$$\left\{\frac{1}{n-2}\right\}_3^\infty = \left\{1, \frac{1}{2}, \frac{1}{3}, ..., \frac{1}{n-2}, ...\right\}. \tag{3.10}$$

The real number S_n is the nth term of the sequence. We take $k = 0$ unless S_n is given by a rule that is invalid for some nonnegative integer, in which case k is understood to be the smallest positive integer such that S_n is defined for all $n \geq k$. For example, if

$S_n = \frac{1}{(n-1)(n-5)}$, then $k = 6$.

3.2.1 Convergence and Divergence of sequences

definition: A sequence $\{a_n\}$ converged to a limit L if for every $\epsilon > 0$ there is an integer N, such that ,

$$\mid a_n - L \mid < \epsilon, \ if \ n \geq N. \tag{3.11}$$

If we can make the terms a_n as close to L as we like by taking n sufficiently large, then we say that $\{a_n\}$ is convergent and write

$$\lim_{n \to \infty} a_n = L \qquad (3.12)$$

In fact if the limit in (3.12) exists, we say the sequence converges (or is convergent), and if the limit in (3.12) does not exist then we say the sequence diverges (or is divergent). Also convergent means the limit on (3.12) must be finite.

For Example: If

$$a_n = \left\{ \frac{2n+3}{n+1} \right\}, \quad then \lim_{n \to \infty} a_n = 3. \qquad (3.13)$$

Since

$$\mid a_n - 3 \mid = \mid \frac{2n+3}{n+1} - 3 \mid = \frac{1}{n+1}. \qquad (3.14)$$

Hence , if $\epsilon > 0$, then equation (3.11) holds with $L = 2$ if $N \geq 1/\epsilon$.

Example: Proof that the sequence ,

$$\{(1+2/n)^n\}_{n=1}^{\infty}, \; is \; convergent. \qquad (3.15)$$

Proof:

Let $S_n = (1+2/n)^n$. Then by the Binomial theorem we can expand this series as,

$$S_n = 1 + n.\frac{1}{n} + \frac{n(n-1)}{1.2}\left(\frac{2}{n}\right)^2 + ... + \frac{n(n-1)...1}{1.2...n}.\left(\frac{2}{n}\right)^n.$$

For the kth term,

$$S_K = \frac{n(n-1)...(n-k+1)}{1.2...k}.\left(\frac{2}{n}\right)^k,$$

And for $(k+1)$st term,

$$S_{K+1} = \frac{n(n-1)...(n-k+2)}{1.2...k} \cdot \left(\frac{2}{n}\right)^{k+1},$$

This proves that $S_n \leq S_{n+1}$ (that is the sequence $\{S_n\}_{n=1}^{\infty}$ is nondecreasing). But also,

$$\left(1+\frac{1}{n}\right)^n = S_n < 1+2+2\frac{1}{1.2}+2\frac{1}{1.2.3}+...+2\frac{1}{1.2...n}$$

$$< 1+2+2.\frac{1}{2}+2.\frac{1}{2^2}+...+2.\frac{1}{2^{n-1}}$$

$$= 1+2.(\frac{1}{2})^0+2.(\frac{1}{2})^1+....2.(\frac{1}{2})^n.$$

$$= 1+\frac{1-(1/2)^n}{1-1/2}$$

$$< 1+\frac{1}{1-1/2} = 3.$$

The underlined is a series with ratio term $r = 1/2$, and first term $a = 1/2$ since $r = 1/2 < 1$ then the series converges, so the sum can be calculated as,

$$S_n = \frac{a(1-r^n)}{1-r} = \frac{2(1-(1/2)^n)}{1-1/2},$$

$$< 1+\frac{2(1-(1/2)^n)}{1-1/2} = 5$$

Hence the sequence $\{S_n\}_{n=1}^{\infty}$ is bounded above, and it is convergent. Let us go back to calculus and use it to proof the convergence of this sequence.

$$Let\ y = \left(1+\frac{2}{x}\right)^x,$$

taking the natural logarithm of both sides,

$$\ln y = x \ln(1 + \frac{2}{x}), \qquad (3.16)$$

Then taking the limit at infinity for both sides,

$$lim_{x \to \infty} \ln y = lim_{x \to \infty} \frac{\ln(1 + \frac{2}{x})}{\frac{1}{x}}. \qquad (3.17)$$

And by using L'Hospital Rule we get,

$$= lim_{x \to \infty} \frac{(\frac{1}{x+2/x})(-2/x^2)}{-1/x},$$

$$= lim_{x \to \infty} \frac{2}{1 + 2/x},$$

$$= 2,$$

$$\Rightarrow lim_{x \to \infty}(1 + \frac{2}{x})^x.$$

Thus, $lim_{x \to \infty} e^{\ln y} = e^2.$

And using the following theorem from Calculus:
If $lim_{x \to} f(x) = L$ and $f(n) = a_n$ when n is an integer, then $lim_{n \to \infty} a_n = L$, we can write, $lim_{n \to \infty}(1 + \frac{2}{n})^n = e^2$ and is convergent.

Note: It is customary to denote $lim_{n \to \infty} S_n$ by e. that is,

$$e = \lim_{n \to \infty} \left(1 + \frac{1}{n}\right)^n. \qquad (3.18)$$

Actually this number e is familiar from calculus, and is a transcendental number whose decimal expansion is 2.718.... We know that the convergent sequence is bounded. Therefore the unbounded sequence is divergent. It seems intuitively obvious that a nondecreasing sequence must diverge to infinity.

This would imply that an unbounded nondecreasing sequence must diverge to infinity.

Theorem: A nondecreasing sequence which is bounded above is convergent.

Definition: Limit of a sequence: Having given an arbitrary small positive number ϵ, if there exist a positive number δ such that;

$| f(x) - L | < \epsilon$, for all values of $x \neq x_0$, which satisfy the condition $| x - x_0 | < \delta$. Then L is said to be the limit of $f(x)$ as x approaches x_0. Symbolically we can write this as, $\lim_{x \to x_0} f(x) = L$.

For every value of x in the interval $(x_0 - \delta, x_0 + \delta)$ the corresponding value of $f(x)$ lies within the rectangle bounded by the lines :

$y = L \pm \epsilon, \quad x = x_0 \pm \delta$

The points representing $f(x)$ are dense in the two dimension neighborhood of L.

Example:

Consider the sequence $\{S_n\}_{n=1}^{\infty}$, where $S_n = (-1)^n$, where, $n = 1, 2, .., n$. The terms of the sequence are:

$$\{S_n\}_{n=1}^{\infty} = \{(-1)^n\} = -1, 1, -1, 1, ... \qquad (3.19)$$

Suppose there is $L \in R$ for which $\lim_{n \to \infty} S_n = L$, then for $\epsilon = 1/2$ there would be $N \in I$ such that (3.19) holds, that is ,

$$| (-1)^n - L | < \frac{1}{2} \qquad (3.20)$$

$$for \ n - even \to | 1 - L | < \frac{1}{2}. \qquad (3.21)$$

$$And \ for \ n - odd \to | -1 - L | < \frac{1}{2}. \qquad (3.22)$$

Since $|a - b| =$ distance from a to b, then (3.21) implies that L is less than $1/2$ unit from 1, while (3.22) implies that L is less than $1/2$ unit away from -1. This is a contradiction.
$\Rightarrow$ The inequality in (3.20) is equivalent to ,

$$|(1 + L)| < \frac{1}{2} \qquad (3.23)$$

$$|-(1 + L)| < \frac{1}{2}. \qquad (3.24)$$

But then,

$$
\begin{aligned}
2 = |2| &= |1 + 1| & (3.25)\\
&= |1 + L + 1 - L| & (3.26)\\
&\leq |1 + L| + |1 - L| & (3.27)\\
&\leq \frac{1}{2} + \frac{1}{2} = 1 & (3.28)\\
& & (3.29)
\end{aligned}
$$

Which is contradiction. Hence, no limit L exist for the sequence $\{S_n\}_{n=1}^{\infty}$, even though the terms of the sequence all have absolute value 1 and hence are not too big.
Note: It is better if we first guess what the limit is if we wish to show that a given sequence has limit.

Example: Let,

$$\{S_n\}_{n=1}^{\infty} = \left\{ \frac{3n}{n + 7n^{1/2}} \right\} \qquad (3.30)$$

When n is large , then it is much larger than $n^{1/2}$, then taking the limit of the sequence our guess would be (if the limit exists) , $lim_{x \to \infty} \{S_n\} = 3$. Then for large n , S_n goes near to 3. We therefore guess that $\{S_n\}_{n=1}^{\infty}$ has the limit of 3.
Now let us prove that $lim_{n \to \infty} S_n = 3$, given that $\epsilon > 0$, we

must find $N \in I$ such that:

$$| S_n - 3 | < \epsilon \, (n \leq N) \qquad (3.31)$$

$$\left| \frac{3n}{n + 7n^{1/2}} - 3 \right| < \epsilon \, (n \leq N) \qquad (3.32)$$

$$\left| \frac{3n - 3n + 21n^{1/2}}{n + 7n^{1/2}} \right| < \epsilon \, (n \leq N) \qquad (3.33)$$

$$\frac{21n^{1/2}}{n + 7n^{1/2}} < \epsilon \, (n \leq N). \qquad (3.34)$$

Since the left side of (3.34) is less than $21n^{1/2}$,

$$\frac{21n^{1/2}}{n + 7n^{1/2}} < \frac{21n^{1/2}}{n} = \frac{21}{n^{1/2}}. \qquad (3.35)$$

Then (3.35) will be true if,

$$\frac{21}{n^{1/2}} < \epsilon, \, (n \leq N) \qquad (3.36)$$

$$\rightarrow \frac{21}{n} < \epsilon^2. \qquad (3.37)$$

Then, we choose $N > \frac{231}{\epsilon^2}$. This proves $\lim_{n \to \infty} s_n = 3$

Theorem:
If $\{s_n\}_{n=1}^{\infty}$ is a sequence of non-negative numbers and if $\lim_{n \to \infty} S_n = L$, then $L \geq 0$.

Proof: Suppose the contrary, namely that $L < 0$, then for $\epsilon = -\frac{L}{2}$ there exist $N \in I$ such that ,

$$| S_n - L | < -\frac{L}{2} \, (n \geq N) \qquad (3.38)$$

$$| S_N - L | < -\frac{L}{2}, \qquad (3.39)$$

$$\Rightarrow S_N - L < -\frac{L}{2}, \qquad (3.40)$$

$$or \, S_N < \frac{L}{2}. \qquad (3.41)$$

But by hypotheses, $S_N \geq 0$, this implies $L > 0$, contradicting our supposition that $L < 0$. Hence $L \geq 0$.

The proof states the following: If S_n gets close to L when n is large, and $L < 0$, then $S_n < 0$ for sufficient large n.

Examples on limits of Algebraic Functions:

1. Prove that,

$$\lim_{x \to 1}(x^2 + 2x) = 3. \tag{3.42}$$

Proof: Using the following,

$$\lim_{x \to x_0} f(x) = L, and \mid f(x) - L \mid < \epsilon, whenever \mid x - x_0 \mid < \delta.$$
$$\tag{3.43}$$

Where $\epsilon > 0$ and $x \neq x_0$.

Now according to (3.42) we have $f(x) = x^2 + 2x$, $L = 3$, $x_0 = 1$. Given $\epsilon > 0$ we must find $\delta > 0$ such that,

$$\mid (x^2 + 2x) - 3 \mid < \epsilon \ (0 < \mid x - 1 \mid < \delta). \tag{3.44}$$

Note that $\mid (x^2 + 2x) - 3 \mid = \mid x - 1 \mid \ . \ \mid x + 3 \mid$. Taking $\mid x - 1 \mid < \delta$, we ask ourselves: how big can $\mid x + 3 \mid$ be ?

Suppose we choose $\delta < 1$. Then,

For $\mid x - 1 \mid < \delta \Rightarrow \mid x - 1 \mid < 1$ Thus $\Rightarrow x \in (0, 2)$ and so $x + 3 \in (5, 7)$.

Hence $\mid x + 3 \mid < 7$ if $\mid x - 1 \mid < \delta < 1$, and so

$\mid x - 1 \mid \ . \ \mid x + 3 \mid < \delta.7$ if $\mid x - 1 \mid < \delta$, and $\delta < 1$.

Let $\delta = min(1, \epsilon/7)$. Thus,

$$\mid x - 1 \mid \ . \ \mid x + 3 \mid < 7 \ \delta \leq \epsilon \quad (\mid x - 1 \mid < \delta), \tag{3.45}$$

Thus (3.42) holds for $\delta = min(1, \epsilon/7)$ for the given $\epsilon > 0$.

2. Show that $\lim_{x \to 1} \sqrt{x+3} = 2$.

Solution: As before applying (3.43) we see that,

$f(x) = \sqrt{x+3}$, $L = 2$, $x_0 = 1$. Given $\epsilon > 0$.
We must find $\delta > 0$ such that,

$$| \sqrt{x+3} - 2 |< \epsilon \quad (0 <| x - 1 |< \delta). \qquad (3.46)$$

Multiplying the first part of equation (3.46) by $| \frac{(\sqrt{x+3}+2)}{(\sqrt{x+3}+2)} |$ gives,

$$\frac{| (\sqrt{x+3})^2 + 2^2 |}{| \sqrt{x+3} + 2 |} < \epsilon, \quad (0 <| x - 1 |< \delta) \qquad (3.47)$$

$$\frac{| x - 1 |}{| \sqrt{x+3} + 2 |} < \epsilon, \quad (0 <| x - 1 |< \delta) \qquad (3.48)$$

If we take $\delta < 1$, then $| x - 1 |< \delta \to x \in (0, 2)$ and hence

$\sqrt{x+3} + 2 > \sqrt{3} + 2$. Thus , if $| x - 1 |< \delta < 1$, then,

$$\frac{| x - 1 |}{| \sqrt{x+3} + 2 |} < \frac{\delta}{\sqrt{3} + 2}, \qquad (3.49)$$

If we pick $\delta = min(1, \epsilon(\sqrt{3} + 2))$ then $\frac{\delta}{\sqrt{3}+2} \le \epsilon$, hence (3.49) holds.

3. Prove that $\lim_{x \to \infty}(1/x^2) = 0$.

Proof: Given $\epsilon > 0$ we must find $M \in R$ such that,

$$| \frac{1}{x^2} - 0 |< \epsilon, \quad (x > M). \qquad (3.50)$$

Since this is equivalent to ,

$$\frac{1}{x} < \sqrt{x}, \quad (x > M), \tag{3.51}$$

then (3.51) hold if we take $M = 1/\sqrt{\epsilon}$.

3.3 Sequence of Real Numbers

Sequence of numbers can be defined (intuitively) as a set of numbers with order.

Definition: A sequence of real numbers $S = s_i{}_{i=1}^{\infty}$ is a function from I into R.

Example: The Fibonacci sequence $\{s_n\}_{n=1}^{\infty}$ given as $S_{n+1} = S_n + S_{n-1}$, where, $n = 2, 3, 4..$, and $S_1 = S_2 = 1$. Find S_8.

Solution:

$$
\begin{aligned}
S_1 &= 1. \\
S_2 &= 1. \\
S_3 &= S_2 + S_1. \\
    &= 1 + 1 = 2. \\
S_4 &= S_3 + S_2 \\
    &= 2 + 1 = 3. \\
S_5 &= S_4 + S_3. \\
    &= 3 + 2 = 5. \\
S_6 &= S_5 + S_4. \\
    &= 5 + 3 = 8. \\
S_7 &= S_6 + S_5. \\
    &= 8 + 5 = 13. \\
S_8 &= S_7 + S_6. \\
    &= 13 + 8 = 21.
\end{aligned}
$$

Monotone Sequence:
Definition: If $\{a_n\}$ is a sequence of real numbers that is monotone and bounded, then it is convergent.

Monotonicity of sequences:

To apply the above theorem we need to know that the sequence is monotone, but sometimes its not easy, we need to analyze the set to determine its monotonicity. The following example shows the steps.
Example: Show that the sequence $\{S_n\} = \frac{2n-1}{n+1}$ is monotone.
Solution: To test the monotonicity of the sequence $\{S_n\}$ we apply the following test:
Test the difference: $\{S_{n+1} - S_n\}$.

$$
\begin{aligned}
S_{n+1} - S_n &= \frac{2(n+1)-1}{(n+1)+1} - \frac{2n-1}{n+1}. \\
&= \frac{2(n+1)[2n+1] - (2n-1)(n+2)}{(n+1)(n+2)}. \\
&= \frac{2(2n^2 + n + 2n + 1) - (2n^2 + 4n - n - 2)}{(n+1)(n+2)}. \\
&= \frac{4n^2 + 6n + 2 - 2n^2 - 3n + 2}{(n+1)(n+2)}. \\
&= \frac{2n^2 + 3n + 4}{(n+1)(n+2)} > 0.
\end{aligned}
$$

Thus for all n, $S_{n+1} - S_n > 0 \rightarrow S_{n+1} > S_n$ or the sequence is increasing.

3.4 Exercise - 1

1. Show that $sin(\frac{1}{x})$ does not approach a limit as $x \to 0$.

2. Prove the truth of the following statements:

$$(a) \; lim_{x \to -2} x^2 + 3x \;\; = \;\; -2. \qquad (3.52)$$

$$(b) \; lim_{x \to 1} \frac{x^2 - 1}{x - 1} \;\; = \;\; 2. \qquad (3.53)$$

$$(c) \; lim_{x \to 0} \sqrt{4 - x} \;\; = \;\; 2. \qquad (3.54)$$

3. Prove that ,

$$\lim_{x \to 1} \frac{x^7 - 2x^5 + 1}{x^3 - 3x^2 + 2} = 1. \qquad (3.55)$$

4. Consider the sequence $1, 1/2, 1/3, ..$, that is, consider $\{S_n\}_{n=1}^{\infty}$ where,

$$S_n = \frac{1}{n} \; (n = 1, 2, ..). \qquad (3.56)$$

Prove that this sequence has a limit $L = 0$.

5. For the Fibonacci numbers the sequence is given as: $\{s_n\}_{n=1}^{\infty}$, and,

$$s_1 \;\; = \;\; 1,$$
$$s_2 \;\; = \;\; 1,$$
$$s_n \;\; = \;\; s_{n-1} + s_{n-2}, \;\; for \; (n \geq 3). \; Find \; s_{10}.$$

6. For each of the following sequences, prove that the sequence has a limit or that the sequence does not have a limit:

$$a) \;\; \left\{ \frac{3n}{n + 7n^{1/2}} \right\}_{n=1}^{\infty}$$

$$b) \;\; \left\{ \frac{n^2}{n + 5} \right\}_{n=1}^{\infty}$$

c) $\left\{\dfrac{3n}{n+7n^2}\right\}_{n=1}^{\infty}$

Chapter 4

Riemann Integral

The integral that is taught in Calculus is called Riemann Integral. It was developed by the German Mathematician Bernhard Riemann, who provided a replacement to the intuitive notion of the integral due to Newton and Leibniz. All the integrals that were defined after Riemann were simply a generalization of his integral. To understand the generalized integrals one needs to understand Riemann's integrals thoroughly.

However the Riemann integral has certain defects which can be remedied by the use of the Lebesgue integral. The main difference between the Riemann integrals and Lebesgue integrals is that Riemann integrals uses intervals and their length while Lebesgue integrals uses more general point sets and their measures. This it is not surprising that the Lebesgue integral is more general than the Riemann integral.

4.1   The Riemann Integral

Integration theories attempts to make the concept of area mathematically precise. Riemann's approach to the integral was modeled on the method of exhaustion used by the ancient Greeks. The scheme is to approximate the area under the graph of a positive function f on the interval $[a, b]$ as follows: Divide $[a, b]$ into subintervals which in turn divides the area under the graph into vertical strips, each of these strips is approximately rectangular. The area under each strip is approximately equal to the area of a rectangle. Finally the area under the graph f between a and b is approximated by adding up the area of all the strips between a and b. Thus, we get the "exact value" for the area under the graph between a and b by means of approximate limiting process.

Definition 4.1.1:

First we have to clarify the meaning of the integral. Riemann's definition for the integral is similar to Cauchy, he uses approximating sums S:

$$S = \int_a^b f(x)dx = \sum_{j=1}^n \left(f(x_{j-1}^*)(x_j - x_{j-1}) \right), \qquad (4.1)$$

where x_{j-1}^* can be any point in the interval $[x_{j-1}, x_j]$; and $1 \le j \le n$.

Definition 4.1.2:
Let $f(x)$ be a function defined on $[a, b]$. We say that $f(x)$ is Riemann integrable on $[a, b]$ if there is a number L with the following properties: For every $\epsilon > 0$ there is a $\delta > 0$ such that:

$$\mid S - L \mid < \epsilon, \qquad (4.2)$$

Which leads to ,

$$\mid \sum f(C_j)(x_j - x_{j-1}) - L \mid < \epsilon, \qquad (4.3)$$

if S is any Riemann sum of f(x) over a partition $p = \{x_0, x_1, ...x_n\}$ of $[a, b]$ such that the norm of $p = \parallel p \parallel = max_{1 \le j \le n}(x_j - x_{j-1}) <$

δ in this case we say that L is Riemann integral of $f(x)$ over $[a, b]$ and write,

$$\int_a^b f(x)dx = L \qquad (4.4)$$

This means the curve between the intervals $[a, b]$ is divided into n-subintervals by points $x_1, x_2, ..., x_n$ where $a = x_0 < x_1 < x_2... < x_n = b$, this is called a partition, net, or mode of subdivisions of the integral, the largest of the values $x_j - x_{j-1} = \| p \|$ where $j = 1, 2, .., n$ is called the norm of the partition and is denoted by δ .

4.2   Construction of the Integral

In calculus students studied integrals of Riemann type. Here we will be revisiting that part from calculus, but with analytical approach, that was not done in calculus. So we will start with the area under the curve divided into subintervals as described in the following section.

4.2.1   Riemann Sum

We understand that finding the integral of a function $f(x)$ inside a closed interval $[a, b]$ means finding the area under the curve of the function $f(x)$ that falls between a, b. This area was formed as follows:

We will divide the interval $[a, b]$ into n-subintervals starting with $a = x_0$, and ending with $b = x_n$, as shown on the line of numbers:

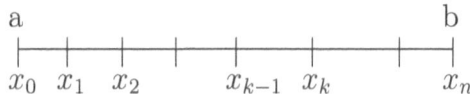

Where, $a = x_0 < x_1 < x_2 \ldots < x_n = b$.

These partitions divide the area under the curve $f(x)$ into blocks each of area= area of a rectangle$\rightarrow$ A = hight x width. Each block goes above and below the curve of $f(x)$, as shown on the first 3-blocks in the graph below.

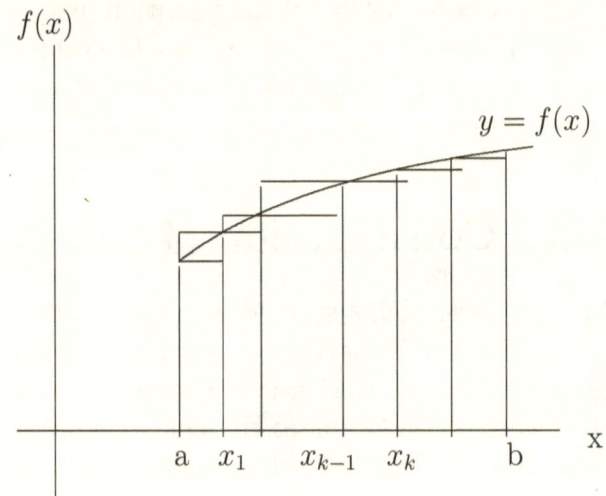

The width of blocks are: $(x_1 - x_0), (x_2 - x_1), \ldots (x_k - x_{k-1})$.

If we use the following notations:

Upper hight $= L_k$.

Lower hight $= G_k$.

Width $= (x_k - x_{k-1}) = \Delta x_k$.

Area of Upper block $= L_k \Delta x_k$.

Area of lower block$= G_k \Delta x_k$.

Sum of area of all upper blocks $= U_s$.

Sum of area of all lower blocks $= D_s$.

Where,

$$U_s = L_1(x_1 - x_0) + L_2(x_2 - x_1) + \ldots + L_k(x_k - x_{k-1}).$$
$$= \sum_{k=1}^{n} L_k \Delta x_k.$$

$$\text{And, } D_s = G_1(x_1 - x_0) + G_2(x_2 - x_1) + ... + G_k(x_k - x_{k-1}).$$

$$= \sum_{k=1}^{n} G_k \Delta x_k.$$

It is obvious from the graph that the sum of the upper area is larger than the lower area. Or $D_s \leq U_s$. As we have mentioned earlier that the area under the curve is the integral of the function $f(x)$ in the finite interval. Then we can use integrals as follows:

$$U_s = \int_a^b f(x)dx = I_U.$$

$$\text{And, } D_s = \int_a^b f(x)dx = I_L.$$

Where, I_U is the integral of the upper area, and I_L is the integral of the lower area. If these two integrals are equal to each other, then we say that $f(x)$ is Riemann Integrable in the closed interval $[a, b]$. We will denote it as I_R:

$$I_R = \int_a^b f(x)dx.$$

If the two integrals are not equal or: $I_U \neq I_L$, then $f(x)$ is not Riemann Integrable on $[a, b]$.

4.2.2 Riemann Integribility Condition

If a function $f(x)$ is a Riemann Integrable in a closed interval $[a, b]$, then it must satisfy the following condition:
For a given $\epsilon > 0$, there exist a subinterval with upper U_s, and lower D_s sums , such that: $U_s - D_s < \epsilon$.

Note: Riemann Integral I_R can be written in limit form such as:

$$\int_a^b f(x)dx = lim_{n \to \infty, \delta \to 0} \sum_{k=1}^{n} f(x_k) \Delta x_k.$$

Where, $\Delta x_k = x_k - x_{k-1}$, and $\delta = max \Delta x_k$.

Theorem: If U_s, D_s are the upper and lower sums to subintervals, and L, G are the upper and lower bounds of $f(x)$ in the interval $[a, b]$, then:

$$\underbrace{G(b-a)}_{\text{lower area}} \leq D_s \leq U_s \leq \underbrace{L(b-a)}_{\text{upper area}}.$$

Riemann Integral Theorems:

1. A continuous function $f(x)$ in $[a, b]$ is Riemann integrable in $[a, b]$.

2. A monotonic function $f(x)$ in $[a, b]$ is Riemann integrable in $[a, b]$.

3. A function $f(x)$ of bounded variation in $[a, b]$ is Riemann integrable in $[a, b]$.

4. A sufficient condition for a bounded function $f(x)$ to be Riemann integrable in $[a, b]$ is to have measure zero of the set of discontinuities at $f(x)$ in $[a, b]$.

Definition: Measure Zero

A set S of real numbers is said to be measure zero if, for $\epsilon > 0$ there exist a countable set of open intervals I_k, $k = 1, 2, ..$ such that $S \subset \cup I_k$, and such that the length of I_k is less than ϵ, i.e: $\sum L(I_k) < \epsilon$. Where $\cup I_k$ is the union of the subintervals I_k.

Mean Value Theorem: If $f(x)$ is continuous in $[a, b]$, then there exist a number $c \epsilon [a, b]$ such that:

$$\int_a^b f(x)dx = (b-a)f(x). \tag{4.5}$$

Theorems:

5. If $f(x)$ is Riemann integrable in $[a, b]$ and $f(x) \leq k$ for some constant k, then

$$\left| \int_a^b f(x)dx \right| \leq k(b - a). \qquad (4.6)$$

6. If $f(x)$ is Riemann Integrable in $[a, b]$, then $|f(x)|$ is Riemann Integrable in $[a, b]$, and

$$\left| \int_a^b f(x)dx \right| \leq \int_a^b |f(x)| \, dx. \qquad (4.7)$$

7. If $f(x)$ is bounded and Riemann Integrable in $[a, b]$, then

$$F(x) = \int_a^x f(t)dt, \qquad (4.8)$$

is continuous in $[a, b]$.

8. If $f(x)$ is Riemann Integrable in $[a, b]$ and $F(x) = \int_a^b f(t)dt$, then

$$F'(x) = \frac{dF(x)}{dx} = \frac{d}{dx} \int_a^b f(t)dt = f(x), \qquad (4.9)$$

at each point of continuity of $f(x)$'

Fundamental Theorem of Calculus:
Let $f(x)$ be Riemann Integrable in $[a, b]$ and suppose that there exist a function $F(x)$ continuous in $[a, b]$ such that,

$$F'(x) = f(x),$$

$$then \int_a^b F'(x)dx = \int_a^b f(x)dx = F(b) - F(a).$$

$$Or \int_a^x F'(t)dt = \int_a^x f(t)dt = F(x) - F(a).$$

Integration by Parts:
Let $f(x)$ and $g(x)$ be Riemann Integrable in $[a, b]$, and suppose
that $F(x)$ and $G(x)$ are such that:

$$F'(x) = f(x).$$
$$And \ G'(x) = g(x) \ in \ [a, b],$$
$$then \int_a^b F(x)g(x)dx = F(b)G(b) - F(a)G(a) - \int_a^b f(x)G(x)dx.$$
$$= F(x)g(x) \ |_a^b - \int_a^b f(x)G(x)dx.$$

Change of Variable Theorem:
Let $f(x)$ be continuous in $[a, b]$, let $x = g(t)$ have a continuous
derivative in (α, β) where, $a = g(\alpha)$, $b = g(\beta)$, then

$$\int_a^b f(x)dx = \int_\alpha^\beta f[g(t)]g'(t)dt.$$

<u>**Examples:**</u>

In the following examples we will show how the accurate approximations for a large class of sums in applied problems, is
the definite integral. To understand it we take $f(x)$ to be continuous on $[a, b]$ and, assume that $f(x) \geq 0$ on $[a, b]$. When
formulating a precise definition, only the continuity of f will
be of importance. The process is to divide the closed interval
$[a, b]$ into n equal subintervals, each of length $| \Delta x | = \frac{b-a}{n}$, by
means of points $a = x_0 < x_1 < x_2 < .... < x_n = b$. Let ξ_i be
a point in the range $x_i \leq \xi_i \leq x_{i+1}$, where $i = 0, 1, ..., n - 1$.
Then we take the sums of the form :

$$S = \sum_{i=0}^{n-1} [f(\xi_i)). \ | \ \Delta x \ | \ ] \qquad (4.10)$$

Example-1:

For a given function $f(x) = x^2$ in $[0, 1]$, where $\mid \Delta x \mid = 0.2$, and $\xi_i = x_i$ for $i = 0, 1, 2, 3, 4$. find the sum value using formula (4.10).

Solution:

First we divide the interval $[0, 1]$ into subintervals, each of width $(.2)$.

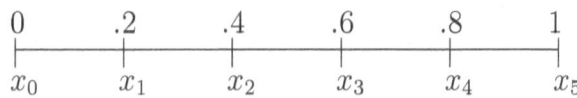

$x_0 = 0.$ $x_1 = 0.2,$ $x_2 = 0.4,$ $x_3 = 0.6,$ $x_4 = 0.8,$ $x_5 = 1,$ and, $f(x_0) = 0,$ $f(x_1) = 0.04,$ $f(x_2) = 0.16,$ $f(x_3) = 0.36,$ $f(x_4) = 0.64,$ $f(x_5) = 1,$ then the sum has the value:

$$S = \sum_{i=0}^{4} [f(x_i). \mid \Delta x \mid], \quad where, \ n - 1 = 4.$$

$S = (0)(0.2) + (0.04)(0.2) + (0.16)(0.2) + (0.36)(0.2) + (0.64)(0.2).$
Then, $\quad S = 0.240.$

Example-2:

On $[0, 1]$, let $f(x) = x^2,$ $\mid \Delta x \mid = 0.1,$ $\xi_i = x_i$ with $i = 0, 1, 2, ..9.$ Find the sum value using formula (4.10).

Solution:

as done in example-1, we divide the interval $[0, 1]$ into subintervals with width $(.1)$.

$$
\begin{array}{ccccccccccc}
0 & .1 & .2 & .3 & .4 & .5 & .6 & .7 & .8 & .9 & 1
\end{array}
$$

$$
x_0 \quad x_1 \quad x_2 \quad x_3 \quad x_4 \quad x_5 \quad x_6 \quad x_7 \quad x_8 \quad x_9 \quad x_{10}
$$

Then the sum has the value:

$$
S = \sum_{i=0}^{9} [f(x_i) . \mid \Delta x \mid], \quad with \ \ n - 1 = 9.
$$

$$
S = (0.1)(0 + 0.01 + 0.04 + 0.09 + 0.16 +
$$
$$
0.25 + 0.36 + 0.49 + 0.64 + 0.81).
$$
$$
Then, \ \ S = 0.285
$$

Suppose we consider the sum of the type,

$$
S = \sum_{i=0}^{n-1} [f(\xi_i) . \mid \Delta x \mid] \tag{4.11}
$$

Which has a relatively large number of terms; that is, n is relatively large. Large number of n results from a problem in astrophysics or atomic physics, it may be so large that even a modern computer can not be applied directly or economically. In cases like this the following technique has proved to be of great value: Instead of considering the given,large n, let n grow beyond all bound, or take $lim_{n \to \infty}$,and approximate (4.11) by,

$$
\int = lim_{n \to \infty} \left[\sum_{i=0}^{n-1} f(\xi_i) . \mid \Delta x \mid \right] . \tag{4.12}
$$

Example-3:

On $[0, 1]$, let $f(x) = x^2$, $| \Delta x | = 10^{-6}$, $\xi_i = x_i$.
Approximate (4.11) by (4.12).
Solution: Divide $[0, 1]$ into n-equal parts, then apply the formula,

$$\sum_{i=0}^{n-1} [f(\xi_i). | \Delta x |] = \sum_{i=0}^{n-1} \left[\left(\frac{i}{n} \right)^2 \left(\frac{1}{n} \right) \right] = \frac{(n-1)(2n-1)}{6n^2}.$$

(4.13)

Thus, from (4.13) we get,

$$\int = lim_{n \to \infty} \frac{(n-1)(2n-1)}{6n^2} = lim_{n \to \infty} \frac{2n^2 - 3n + 1}{6n^2} = \frac{1}{3}.$$

(4.14)

Example-4:
For a given function $f(x)x^2 - 2x$ on the closed interval $[-1, 3]$.
Using the following point of division: $x_0 = -1, x_1 = -.2$,
$x_2 = .6, x_3 = 1.8, x_4 = 2.4, x_5 = 3$, find the sum value using
the following formula:

$$S = \sum_{k=1}^{n} [f(C_k). | \Delta x_K |].$$

(4.15)

$$Where, \quad C_k = \frac{x_k + x_{k-1}}{2}.$$

(4.16)

$$And, \quad \Delta x_k = x_k - x_{k-1}.$$

(4.17)

Solution:
Using (4.16) we can find the $C_i's$:

$$c_1 = \frac{-1 - .2}{2} = -.6.$$

$$c_2 = \frac{.6 - .2}{2} = .2.$$

$$c_3 = \frac{1.8 + .2}{2} = 1.2.$$

$$c_4 = \frac{2.4 + 1.8}{2} = 2.1.$$

$$c_5 = \frac{3 + 2.4}{2} = 2.7.$$

And using (4.17) we find Δx_k:

$$\Delta x_1 = -.2 + 1 = .8.$$
$$\Delta x_2 = .6 + .2 = .8.$$
$$\Delta x_3 = 1.8 - .6 = 1.2.$$
$$\Delta x_4 = 2.4 - 1.8 = .6.$$
$$\Delta x_5 = 3 - 2.4 = .8.$$

The maximum value of $\Delta x_k = 1.2$. Then by formula (4.15) the sum value is:

$$S = \sum_{k=1}^{n} [f(C_k). \mid \Delta x_K \mid].$$

$$= \sum_{k=1}^{5} [f(C_k). \mid \Delta x_K \mid].$$

$$= f(c_1)\Delta x_1 + f(c_2)\Delta x_2 + f(c_3)\Delta x_3 +$$
$$f(c_4)\Delta x_4 + f(c_5)\Delta x_5.$$
$$= (-.84)(.8) + (.44)(.8) + (3.84)(1.2) +$$
$$(8.61)(.6) + (12.69)(.6).$$
$$Then, \ S = 17.068.$$

4.2.3   Properties of Riemann Integrable Functions

1. If $f(x)$ and $g(x)$ are Riemann integrable in $[a, b]$, then $f(x) + g(x)$ is Riemann integrable in $[a, b]$:

$$\int_a^b [f(x) + g(x)]dx = \int_a^b f(x)dx + \int_a^b g(x)dx \qquad (4.18)$$

2. If $f(x)$ is Riemann integrable in $[a, b]$ and c is any constant , then $cf(x)$ is Riemann integrable in $[a, b]$,

$$\int_a^b cf(x)dx = c \int_a^b f(x)dx \qquad (4.19)$$

3. If $f(x)$ and $g(x)$ are Riemann integrable in $[a, b]$, then $f(x)g(x)$ is Riemann integrable in $[a, b]$

4. If $f(x)$ is Riemann integrable in $[a, b]$, then

$$\int_a^b f(x)dx = - \int_a^b f(x)dx, \qquad (4.20)$$

$$\int_a^b f(x)dx = 0. \qquad (4.21)$$

5. If $f(x)$ is bounded and Riemann integrable in $[a, b]$ and c is any positive point of $[a, b]$ then,

$$\int_a^b f(x)dx = \int_a^c f(x)dx + \int_c^b f(x)dx \qquad (4.22)$$

6. If $f(x)$ and $g(x)$ are Riemann integrable in $[a, b]$ and $f(x) \leq g(x)$, then

$$\int_a^b f(x)dx \leq \int_a^b g(x)dx \qquad (4.23)$$

7. If $f(x)$ is Riemann integrable in $[a, b]$ and has upper bound and lower bound M and m in $[a, b]$ respectively, then

$$m(b - a) \leq \int_a^b f(x)dx \leq M(b - a) \qquad (4.24)$$

8. Mean Value Theorem: If $f(x)$ is continuous in $[a, b]$, then there exist a number $c \in [a, b]$ such that,

$$\int_a^b f(x)dx = (b - a)f(c) \qquad (4.25)$$

9. If $f(x)$ is Riemann integrable in $[a, b]$ and $f(x) \leq M$ for some constant M, then,

$$| \int_a^b f(x)dx | \leq M(b - a) \qquad (4.26)$$

10. If $f(x)$ is Riemann integrable in $[a, b]$, then $f(x)$ is Riemann integrable in $[a, b]$ and,

$$| \int_a^b f(x)dx | \leq \int_a^b f(x)dx \qquad (4.27)$$

4.2.4   Important Riemann Integrals

If $f(x)$ becomes infinite at one or more points of $[a, b]$ then Riemann integral is called improper integral. Such integrals are defined by appropriate limitation:

$$\textbf{1.} \int_a^\infty f(x)dx = lim_{b \to \infty} \int_a^b f(x)dx \qquad (4.28)$$

$$\textbf{2.} \int_{-\infty}^b f(x)dx = lim_{a \to -\infty} \int_a^b f(x)dx \qquad (4.29)$$

$$\textbf{3.} \int_{-\infty}^\infty f(x)dx = lim_{a \to -\infty \, b \to \infty} \int_a^b f(x)dx. \qquad (4.30)$$

Problems:

1. prove that the function,

$$f(x) = \begin{cases} 1 \ if \ x \ rational \\ 0 \ if \ x \ irrational \end{cases}$$

Where $f(x)$ is defined in $[a, b]$, is not Riemann integrable.

Proof:
Let the upper bound of $f(x)$ is $M_i = 1$.
And the lower bound of $f(x)$ is $m_i = 0$.
Then the upper sum $= S = \sum_{i=1}^{n} M_i \Delta x_i = \sum_{i=1}^{n} 1.\Delta x_i = b - a$
And the lower sum $= s = \sum_{i=1}^{n} m_i \Delta x_i = \sum_{i=1}^{n} 0.\Delta x_i = 0$
Thus $I = b - a$, and $J = 0$. Since $I \neq J$ then $f(x)$ is not
Riemann Integrable.

2. If S, s are the upper and lower sums corresponding to partition p, and M, m are the upper and lower bounds of $f(x)$ in $[a, b]$, then,

$$m(b - a) \leq s \leq S \leq M(b - a) \tag{4.31}$$

Proof:

$$S = \sum_{i=1}^{n} M_i \Delta x_i \tag{4.32}$$

$$s = \sum_{i=1}^{n} m_i \Delta x_i. \tag{4.33}$$

Since $m \leq m_i \leq M_i \leq M$. Multiply both sides by Δx_i, and summing over $i, i = 1, 2, .., n$, we get,

$$\sum_{i=1}^{n} m \Delta x_i \leq \sum_{i=1}^{n} m_i \Delta x_i \leq \sum_{i=1}^{n} M_i \Delta x_i \leq \sum_{i=1}^{n} M \Delta x_i, \tag{4.34}$$

Using the definition of S, s, we get,

$$\sum_{i=1}^{n} m \Delta x_i \leq s \leq S \leq \sum_{i=1}^{n} M \Delta x_i, \tag{4.35}$$

$$m \sum_{i=1}^{n} \Delta x_i \leq s \leq S \leq M \sum_{i=1}^{n} \Delta x_i, \tag{4.36}$$

$$m(b-a) \leq s \leq S \leq M(b-a). \qquad (4.37)$$

3. For $\epsilon > 0$, if we choose $\delta > 0$ such that $x_{i-1} \leq \xi_i \leq x_i$, then,

$$| \sum_{i=1}^{n} f(\xi_i) \Delta x_i - \int_a^b f(x) dx \ |< \epsilon \qquad (4.38)$$

Whenever $\Delta x_i \leq \delta$.

Proof:

Let S, s be the upper and lower sum corresponding to a given partition. Then since $m_i \leq f(\xi_i) \leq M_i$, where M_i, m_i are the upper and lower bounds of $f(x)$ in (x_{i-1}, x_i) we have,

$$s \leq \sum_{i=1}^{n} f(\xi_i) \Delta x_i \leq S, \qquad (4.39)$$

we also have,

$$S \geq \int_a^b f(x) dx \geq s, \qquad (4.40)$$

If we switch the inequality,

$$-S \leq - \int_a^b f(x) dx \leq -s, \qquad (4.41)$$

Adding equations (4.54) and (4.56) gives,

$$s - S \leq \sum_{i=1}^{n} f(\xi_i) \Delta x_i - \int_a^b f(x) dx \leq S - s, \qquad (4.42)$$

Since $S - s > 0$, and $S - s < \epsilon$, then

$$-(S - s) \leq \sum_{i=1}^{n} f(\xi_i) \Delta x_i - \int_a^b f(x) dx \leq S - s, \qquad (4.43)$$

$$| \sum_{i=1}^{n} f(\xi_i) \Delta x_i - \int_a^b f(x) dx \ |\leq S - s, \qquad (4.44)$$

$$\mid \sum_{i=1}^{n} f(\xi_i)\Delta x_i - \int_a^b f(x)dx \mid \leq \epsilon. \qquad (4.45)$$

Theorem:
A continuous function $f(x)$ in $[a, b]$ is Riemann integrable in $[a, b]$.

proof :

Since $f(x)$ is continuous in the closed intervals $[a, b]$, it is uniformly continuous. If we choose any two-points say x_1, x_2 in the interval (x_{i-1}, x_i) then for a given $\epsilon > 0$, there exist $\delta > 0$ such that:

$$\mid f(x_1) - f(x_2) \mid < \frac{\epsilon}{b-a}, whenever \mid x_1 - x_2 \mid < \delta. \qquad (4.46)$$

Thus we can choose points of such intervals so that:

$$M_i - m_i < \frac{\epsilon}{b-a}, \qquad (4.47)$$

taking the upper and lower sums corresponding to the partition of $f(x)$ in $(x_{i-1} - x_i)$ given as,

$$S = \sum_{i=1}^{n} M_i \Delta x_i, \quad s = \sum_{i=1}^{n} m_i \Delta x_i, \qquad (4.48)$$

taking the difference between the sums gives,

$$S - s = \sum_{i=1}^{n}(M_i - m_i)\Delta x_i < \sum_{i=1}^{n} \frac{\epsilon}{b-a}\Delta x_i = \epsilon. \qquad (4.49)$$

Thus $S - s < \epsilon$, then but follows that $f(x)$ is Riemann integrable. QED.

Theorem:
A monotonic function $f(x)$ in $[a, b]$ is Riemann integrable in $[a, b]$.

proof :

Suppose $f(x)$ is monotonic increasing , this means the
partition $a = x_0 < x_1 < ... < x_n = b$; and
$f(a) = f(x_0) \leq f(x_1) \leq ... \leq f(x_n) = f(b)$.
Then it is clear that :

$$m_i = f(x_{i-1}) \tag{4.50}$$
$$M_i = f(x_i). \tag{4.51}$$

Taking the difference between upper and lower sums gives,

$$S - s = \sum_{i=1}^{n} | f(x_i) - f(x_{i-1}) | \Delta x_i, \tag{4.52}$$

assuming that $f(b) \neq f(a)$, $\Delta x_i < \frac{\epsilon}{f(b)-f(a)}$; $f(x_i) \geq f(x_{i-1})$,
then,

$$S - s < \frac{\epsilon}{f(b) - f(a)} \sum_{i=1}^{n} | f(x_i) - f(x_{i-1}) | \tag{4.53}$$
$$= \frac{\epsilon}{f(b) - f(a)} | f(b) - f(a) | = \epsilon \quad QED. \tag{4.54}$$

4.3   Exercise - 1

1.   Find the Riemann sum for the given function on the
given interval and corresponding to the given division:

a. $f(x) = 10(x - 3)$ on$[-1, 3]$; $x_0 = -1, x_1 = 0, x_2 = 3/2, x_3 = 3$;
   $c_1 = -1/2, c_2 = 1/2, c_3 = 2$.

b. $f(x) = 3 - 2x$ on$[0, 3]$; $x_0 = 0, x_1 = 1, x_2 = 3$;
   $c_1 = 1/2, c_2 = 3$.

Chapter 5

Metric Space and Limits

A metric space is a generalization of a Euclidean Space. A Euclidean space is defined by a Cartesian product $R^n = R \times R \times R.. \times R[n - times]$, and this is called n-dimensional Euclidean space . A point in this n-dimensional Euclidean space is an ordered n-tuples $(x_1, x_2, ...x_n)$ of real numbers. If $x = (x_1, x_2, ..., x_n)$, $y = (y_1, y_2, ..., y_n)$ the Euclidean distance between x and y is defined by,

$$d(x, y) = \sqrt{(x_1 - y_1)^2 + (x_2 - y_2)^2 + ... + (x_n - y_n)^2}. \quad (5.1)$$

For $n = 1$, $d(x, y) = \mid x - y \mid$.
Where, from the absolute value theorem,

$$\mid a \mid = \begin{cases} +a \ if \ a > 0 \\ -a \ if \ a < 0 \end{cases}$$

The set of points $\{x : d(x, y) < r\}$ is called an open sphere with the center at r, and the set of points $\{x : d(x, y) \leq r\}$ is closed sphere. When $n = 1$ then it is called the set of open and closed intervals

The distance formula for R^2, R^3, and R^n respectively can be written as,

$$d\left((x_1, x_2), (y_1, y_2)\right) = \sqrt{(x_1 - y_1)^2 + (x_2 - y_2)^2},$$

$$d\left((x_1, x_2, x_3), (y_1, y_2, y_3)\right) = \sqrt{(x_1 - y_1)^2 + (x_2 - y_2)^2 + (x_3 - y_3)^2},$$

$$d\left((x_1, x_2, ..., x_n), (y_1, y_2, ..., y_n)\right) = \\ \sqrt{(x_1 - y_1)^2 + (x_2 - y_2)^2 + ... + (x_n - y_n)^2}.$$

Examples of Metric Spaces:

1. The absolute value metric: the function d is defined by $d(x, y) = | \, x - y \, |$, and is a metric for the set R. The resulting metric space $\langle R, d \rangle$ is denoted by R^1.

2. The discrete metric $d(x, y)$: For the set R, the metric d: $R \times R \rightarrow [0, \infty)$ is defined by,

$$\begin{aligned} d(x, y) &= 0, \quad (x \in R), & (5.2) \\ d(x, y) &= 1, \quad (x, y \in R; x \neq y). & (5.3) \end{aligned}$$

The resulting metric space $\langle R, d \rangle$ is denoted by R_d.

3. The Euclidean metric: The metric $d(x, y)$ for $n = 2$ is the usual distance formula for points in Cartesian plane. Here d satisfies the triangle inequality property, and for two ordered n-tuples of real numbers is defined by,

$$d(x, y) = \left[\sum_{k=1}^{n} (x_k - y_k)^2 \right]^{1/2}. \qquad (5.4)$$

The metric space formed of all n-tuples of real numbers with this metric d is denoted by R^n.

4. The metric of set of all bounded sequences of real numbers defined by,

$$d(x, y) = l.u.b. \mid x_n - y_n \mid . \tag{5.5}$$

Where, $x = \{x_n\}_{n=1}^{\infty}$, and $y = \{y_n\}_{n=1}^{\infty}$ are points in the metric space $\{l^\infty, d\}$ which is denoted by l^∞. d satisfies the first three requirements for the metric.

5. For $x, y \in l^2$ defined by $d(x, y) = \| x - y \|_2$. d is the metric for l^2. the resulting metric space (l^2, d) is denoted by l^2.

6. A distance between two functions $f(x)$ and $g(x)$ in L^p defined by,

$$d(f, g) = \| f - g \| = \left\{ \int_a^b \mid f(x) - g(x) \mid^p dx \right\}^{1/p} \tag{5.6}$$

Is a metric space, also called the normed metric space with triangle inequality: $\| f + g \| \le \| f \| + \| g \|$. The metric space is not the set X of its points. Since it is in fact the pair (x, d) consisting of the set of its points together with the metric d. For example the set of all n-tuples of real numbers can also be made into a metric space by use of the metric d^* given by : $d^*(x, y) = \mid x_1 - y_1 \mid + ... + \mid x_n - y_n \mid$, and this is not the metric space $R^n (if n > 1)$.

5.1 Measuring Distances

Since analysis is known to be the study of closeness, Thus we need to study the distance in order to define closeness. Distance can be measured between all sorts of mathematical objects such as distance between two points in the plane, between two sphere in 3D, or between two functions.

Definition: A metric or distance function for a set of elements S is a function, say $d(x, y)$, on the set of all points x, y of elements of S to the real continuum.

For $x, y \in R$, the geometric interpretation of $\mid x - y \mid$ is the distance from x to y. If we define distance function d by,

$$d(x, y) = \mid x - y \mid, \ (x, y \in R), \tag{5.7}$$

And if for any points $x, y, z \in R$, the following properties hold:

1. The distance from a point to itself is 0: $d(x, x) = 0$, (zero property).

2. The distance between two distinct points is positive: $d(x, y) > 0, \quad x \neq y,,$ (non-negative property).

3. The distance from x to y is equal to the distance from y to x: $d(x, y) = d(y, x)$, (symmetric property)

4. $d(x, y) \leq d(x, z) + d(z, y)$, (Triangle inequality).

Then the distance function d in (5.7) is called Metric.

Definition: Let A be any set with the function d satisfying the above four properties as follows:

1. $d(x, x) = 0, \ (x \in A)$.

2. $d(x, y) > 0, \ (x \neq y, \ x, y \in A)$.

3. $d(x, y) = d(y, x), \ (x, y \in A)$

4. $d(x, y) \leq d(x, z) + d(z, y)$, $(x, y, z \in A)$,   (Triangle inequality).

If d is metric for A, then the ordered pair $\langle A, d \rangle$ is called a metric space. The metric for A thus has the properties $(1) - (4)$ of the distance $\mid x - y \mid$ for R.
Here we will list some important definitions.

Definition-1: A set of elements is called Hausdorff Space if with each element of the set there is associated a class of subsets (of the given set) known as neighborhoods, which satisfy the following postulate:
1. Every element x has at least one neighborhood N_x and is an element of each of its neighborhoods.
2. If there exist two neighborhoods of x, then there exists a neighborhood of x which is contained in each of these.
3. If y is an element of an N_x, there exist a neighborhood N_y of y which is contained in N_x.
4. If x and y are distinct, there exist an N_x and an N_y with no common elements.

Definition-2: A set with metric is called a metric space.

Definition-3: Let x_0 be an element of metric space S, and let r be a positive number. The subset $B(x_0, r)$ of S which consists of all those elements x of S for which $d(x_0, x) < r$ is called the open sphere with center x_0 and radius r.

Theorem: Every metric space can be made into a Hausdorff Space in which the open spheres constitute the set of neighborhood for the respective centers.
Proof: Let x_0 be the element of the given metric space. We define a neighborhood of x_0 as any open sphere with center at x_0. Then we will show that the set of all open spheres satisfies

the postulates for neighborhoods.

1. Every element x has an open sphere and is an element of it. This follows from definition-3.

2. If there are two open spheres of x_0 say $B(x_0, r_1)$, and $B(x_0, r_2)$ and if $r_1 \leq r_2$, then $B(x_0, r_1)$ is in both spheres; if $r_2 \leq r_1$, then $B(x_0, r_2)$ is in both.

3. If y lies in $B(x_0, r_0)$, let $r = r_0 - d(x_0, y)$. Let x be any point in $B(y, r)$. Then the triangle property,
$$d(x_0, x) \leq d(x_0, y) + d(y, x) < d(x_0, y) + r_0 - d(x_0, y) = r_0$$
Hence any element in $B(y, r)$ is contained in $B(x_0, r_0)$, and $B(y, r) \subseteq B(x_0, r_0)$.

4. If $x_0 \neq y_0$, $d(x_0, y_0) = r > 0$.

The two open spheres $B(x_0, r/3)$, and $B(y_0, r/3)$ have no points in common. Suppose X were an element of both spheres, then $d(x_0, x) < r/3$, and $d(y_0, x) < r/3$. But
$$d(x_0, y_0) = r \leq d(x_0, x) + d(x, y_0) < r/3 + r/3 = 2r/3.$$
Which is impossible. This completes the proof of the theorem.

Exercise:

1. Show that if d is a metric for A then so is $2d$.

2. Show that if d and σ are both metrics for a set A, then $d + \sigma$ is also a metric for A.

3. Let l^1 be the class of all sequences $\{s_n\}_{n=1}^{\infty}$ of real numbers such that $\sum_{n=1}^{\infty} | s_n | < \infty$. Show that if $s = \{s_n\}_{n=1}^{\infty}$ and $t = \{t_n\}_{n=1}^{\infty}$ are in l^1, then $d(s, t) = \sum_{n=1}^{\infty} | s_n - t_n |$ defines a metric for l^1.

4. Let X be any non-empty set. for $a, b \in X$, define

$$d(a, b) = \begin{cases} 0 \ if \ a = b \\ 1 \ if \ a \neq b \end{cases}$$

Verify that (X, d) is a metric space.

Note: this metric is called the trivial metric.

5.2   limits in Metric Spaces

Recall from calculus the definition of limit of a function of real variables is based on definition of continuous functions and derivatives.

Definition:

Let $\epsilon \in R$, and let f be a real valued function whose domain includes all points in some open intervals $(a - h, a + h)$ except possibly the point a itself. Then $f(x)$ approaches L (where $L \in R$) as x approaches a if for a given $\epsilon > 0$ there exist $\delta > 0$ such that,

$$| f(x) - L | < \epsilon, \quad (0 <| x - a |< \delta). \tag{5.8}$$

And we write,

$$\lim_{x \to a} f(x) = L \ \ or \ \ f(x) \to L \ \ as \ \ x \to a. \tag{5.9}$$

Note: we emphasize that the point a need not be in the domain of f. Remember the example from Calculus : $\lim_{x \to 0}(sinx/x) = 1$, even though $sinx/x$ is not defined for $x = 0$.

Theorem: Let (M, d) be a metric space and let a be a point in M. Let f and g be real-valued function(which means a function with range in R^1, and the metric in the range is the absolute-value metric) , whose domains are subsets of M.
If $\lim_{x \to a} f(x) = L$ and $\lim_{x \to a} g(x) = N$, then

$$\lim_{x \to a}[f(x) + g(x)] \ = \ L + N, \tag{5.10}$$

$$\lim_{x \to a}[f(x) - g(x)] \ = \ L - N, \tag{5.11}$$

$$\lim_{x \to a}[f(x)g(x)] \ = \ LN, \tag{5.12}$$

and, if $N \neq 0$,

$$\lim_{x \to a} \frac{f(x)}{g(x)} = \frac{L}{N}. \tag{5.13}$$

Proof: Since $\lim_{x \to a} g(x) = N$, we have, for some $\delta_1 > 0$,

$$|g(x) - N| < 1, \quad (0 < d(x,a) < \delta_1).$$
$$\text{Thus, } |g(x)| < |N| + 1 = Q, \quad (0 < d(x,a) < \delta_1).$$
$$\text{Now, } f(x)g(x) - LN = f(x)g(x) - Lg(x) + Lg(x) - LN$$
$$= g(x)[f(x) - L] + L[g(x) - N],$$
$$|f(x)g(x) - LN| \leq |g(x)| \cdot |f(x) - L| + |L| \cdot |g(x) - N|.$$

Hence, if $0 < d(x,a) < \delta_1$,

$$|f(x)g(x) - LN| \leq Q \cdot |f(x) - L| + |L| \cdot |g(x) - N|. \tag{5.14}$$

Given, $\epsilon > 0$ there exist $\delta_2 > 0$ such that

$$Q\,|f(x) - L| < \frac{\epsilon}{2}, \quad (0 < d(x,a) < \delta_2), \tag{5.15}$$

if we let $\delta = min(\delta_1, \delta_2)$, then from the above inequalities we get:

$$|f(x)g(x) - LN| < \epsilon, \quad (0 < d(x,a) < \delta). \tag{5.16}$$

This proves $\lim_{x \to a} f(x)g(x) = LN$.
In the same manner one can proof the rest.

5.3 Some Theorems from Calculus

In this section we include some proofs that might be helpful for the student to go over, as a review for the material studied in Calculus-II, and assistance for Vector Calculus.

The Sandwich Theorem :

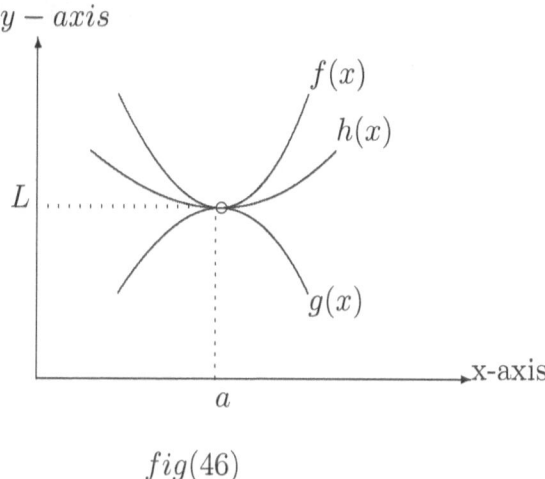

$$fig(46)$$

Suppose $h(x)$ is a function located as shown in fig(46), such as: $f(x) \le h(x) \le g(x)$, and for all $x \ne a$,

$$lin_{x \to a} f(x) = lim_{x \to a} = L, \tag{5.17}$$

then: $lim_{x \to a} h(x) = L.$

$$then: \quad lim_{x \to a} h(x) = L. \tag{5.18}$$

Proof

Assuming that the 2-sided limits of both functions $f(x)$, and $g(x)$ equal some constant L as x approaches a, this means, for some $\in > 0$ there exists $\delta > 0$ such that:

$$L - \in < f(x) \quad \le \quad h(x) \le g(x) < L + \in .$$

$$Then, L- \in \;\; < \;\; h(x) < L+ \in .$$
$$Then \; for \; all \; x: \;\; c < x < c+\delta \;\; \Rightarrow \;\; \|h(x) - L\| < \in .$$
$$then: \;\; lim_{x \to c+} h(x) = L.$$

In a similar way it can be shown that : $lim_{x \to c-} h(x) = L$.
If the 2-sided limits for $h(x)$ approach L, then equation (4.32)
is satisfied.

The Zipper Theorem

If the sequence $\{a_n\}$, and $\{b_n\}$ both converge to L, then the
sequence, $a_1, b_1, a_2, b_2, \cdots, \cdots, a_n, b_n$, also converges to L.
The proof can be done in a way similar to the squeeze theorem.

The Divergence Theorem of Gauss

The Divergence theorem states that: The surface Integral of
the normal component of a vector taken over a closed surface
is equal to the integral of the divergence of the vector over the
volume enclosed by the surface:

$$\int \int_S F.ds = \int \int_S F.nds = \int \int \int_V \nabla .F dV. \qquad (5.19)$$

Where n is the positive (outward) normal to S. The divergence
Theorem is a generalization of Green's Theorem.

Stokes Theorem

Stokes Theorem states that: The line integral of the tangential
component of a vector taken around a simple closed curve is
equal to the surface integral of the normal component of the
curl of the vector taken over any surface having the curve as
its boundary.

$$\oint_C F.dr = \int \int_S (\nabla \times F).ndS = \int \int_S (\nabla \times F).dS. \quad (5.20)$$

Where C in counterclockwise direction.

Green's Theorem in the plane

Green's Theorem relates a closed region in R to a closed curve that bounds the region.

$$\oint_C Mdx + Ndy = \int \int_R (\frac{\partial N}{\partial x} - \frac{\partial M}{\partial y})dxdy. \quad (5.21)$$

Green's Theorem in the plane is a special case of Stokes Theorem.

Notice how Greens Theorem is related to both Divergence Theorem:

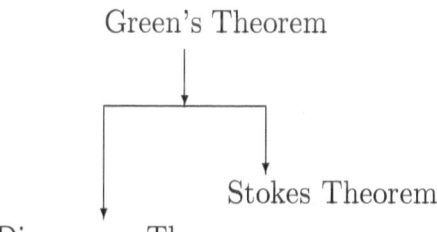

Green's Theorem

Stokes Theorem

Divergence Theorem

Index

A

Bibliography

[1] Carol Schumacher, Closer and Closer Introducing Real Analysis, Jones and Bartlett Publishing , (2008).

[2] David A. Sprecher, Elements of Real Analysis, Academic Press, New York and London , (1970).

[3] David M. Bressoud, A Radical Approach to real Analysis, The Mathematical Association of America, (1991).

[4] E. J. Townsend, Functions of Real Variables, Henry Holt and Company, New York, (1928).

[5] John J. Herriot, Methods of Mathematical Analysis and Computation, John Wiley and Sons, Inc. New York. London , (1963).

[6] H. L. Royden, Real Analysis. The Macmillan Company, New York , (1963).

[7] Henry P. Thielman, Theory of Functions of real variables, Prentice-Hall, Inc., England Cliffs, N. J., (1953).

[8] James Pierpont, The Theory of Functions of Real Variables, Volume-1, Ginn and Company , (1905).

[9] James Stewart, Single Variable Calculus, Thomson-Brooks/Cole,(2005).

[10] Murray R. Spiegel, Theory and Problems of Real Variables, Schaum's Outline Series, McGraw Hill Book Company, (1969).

[11] P. C. Chakravarti, Integrals and Sums, The Athlone Press , (1970).

[12] Richard R. Goldberg, Methods of Real Analysis, Blaisdell Publishing Company, ((1954).

[13] Stanley I. Grossman, Multi variable Calculus, Harcourt Brace Jovanovich, publishers,(1986).

[14] William F. Trench, Introduction to real Analysis, Library of Congress Cataloging - in - Publication Data, (2003).

[15] Yu V. Vorobyev, Methods of Moment in Applied Mathematics, Gordon and Breach Publishers, (1965).

www.ingramcontent.com/pod-product-compliance
Lightning Source LLC
Chambersburg PA
CBHW030751180526
45163CB00003B/983